THE COMPLETE GUIDE TO
Being an
Independent
Contractor

P9-ARD-481

Herman Holtz

Upstart Publishing Company, Inc.
a division of Dearborn Publishing Group, Inc.

While a great deal of care has been taken to provide accurate and current information, the ideas, suggestions, general principles and conclusions presented in this text are subject to local, state and federal laws and regulations, court cases and any revisions of same. The reader is thus urged to consult legal counsel regarding any points of law—this publication should not be used as a substitute for competent legal advice.

Publisher: Anita A. Constant
Executive Editor: Bobbye Middendorf
Managing Editor: Jack L. Kiburz
Editorial Assistant: Stephanie Schmidt
Interior Design: Lucy Jenkins
Cover Design: The Publishing Services Group

Published by Upstart,
a division of Dearborn Publishing Group, Inc.

Printed in the United States of America

95 96 97 10 9 8 7 6 5 4 3 2

Library of Congress Cataloging-in-Publication Data

Holtz, Herman.
   The complete guide to being an independent contractor / Herman
Holtz.
   p. cm.
   Includes bibliographical references and index.
   ISBN 0-936894-80-6 : $24.95
   1. Independent contractors. 2. Self-employed. 3. New business
enterprises—Management. I. Title.
HD2365.H65 1994
658′.041—dc20   94-26905
                                                 CIP

Contents

Appendix: Other Resources   **227**

Index   **241**

Introduction
Why Independent Contracting?

Professional entertainers have been known to say, "Dying is easy; comedy is hard." Business is much like that: Dying (failing) is easy; success is hard. There probably are no truly "easy" businesses, but independent contracting is one of the easier ones to launch and, in many respects, is one of the easier ones in which to find success. Among its several virtues as a business venture are these:

- Independent contracting usually requires little or no capital to launch.
- It can often be conducted from your private home (at least in the beginning).
- It offers tax benefits when operated from your home.
- It is a flexible kind of business, making it relatively easy to expand.

We are in an era of great social and economic change. The most recent recession, allegedly at the end now, was responsible for many layoffs. It also triggered many permanent terminations of jobs. These were different from the classic industry layoffs. In the past, workers eventually were called back from layoffs. The unfortunate individuals whose jobs were eliminated in the recent recession were not merely laid off.

These terminations would have come about, regardless of the recession, because of inevitable change in modern history. They were part of an economic trend that started some time ago and that most of us probably did not recognize.

Economists and social philosophers have long observed the U.S. economy's shift from being predominantly industrial/manufacturing to becoming predominantly service-oriented. That may be bad news for many of the nation's industrialists, the operators of our industrial supercorporations. Still, it is not an unconditional truth, since some of our industries—automobile manufacturing, for example—appear to be making a comeback after losing much of their initiative to other nations. However, it is good news for millions of us known as "opportunity seekers," and others who are ambitious to launch their own independent business ventures, based in their own homes.

This change is a continuing trend. The invention of the digital computer spawned what soon became known as the "information industry." The real revolution came with the introduction of the microchip in computers, which immediately challenged the mainframe computer. It also inspired an explosion in technological development and automation.

Modern automation is changing the nature of the workplace in many establishments, so that fewer hands are needed. Those same forces also are encouraging a dynamic trend to home-based businesses and in general to independent small enterprises. Many people perceive these developments as an unprecedented opportunity for self-realization: Equipped with computers, modems, fax machines and laser printers, their own homes become equal functionally to the facilities of the largest corporations. It has been publicized as the *home office,* but that term does not reflect the true nature of what is happening. What is happening is a revolutionary development of home-based businesses, an apparent harbinger of a vastly greater and more effective cottage industry—a new industrial revolution.

One of the many popular home-based industries—or perhaps it is a mode of launching and conducting a small business, rather than an industry in itself—is independent contracting. Under today's conditions, many employers find it more efficient to contract work to independents and minimize their staffs than to continue supporting a large permanent payroll.

Independent contractor is a rather flexible term, including an extremely wide range of entrepreneurial options for any individual who wishes to embark on an independent business venture. The enormous variety of kinds of work that can be done under contract is as much in the interests of clients as it is in the interests of independent entrepreneurs. Many business organizations find it advantageous to contract for services and functions, especially when it is not clear whether the need is a permanent one. Some businesses, in fact, rely on independents, as in the case of retailers who must provide installation service for their customers of draperies, floor tile and appliances. Aside even from that, employers are increasingly contracting for temporary help—"contract labor," as some temporary services put it—to provide a wide variety of occupational specialties, including technical and professional functions. Many individuals have begun contracting of their services while looking for a permanent employment, while others have discovered advantages in selling their services as contracted-for packages, and have opted for this over permanent in-house employment.

This book offers a road map for many individuals, including those who have lost jobs and are seeking alternatives to regular employment, and those who are seeking guidance in launching a successful independent enterprise.

Technically, *service* businesses are *contracting* businesses, even for such small businesses as beauty parlors, dry-cleaning establishments and copy shops. These, however, are not in the category of what we generally regard as independent contracting. That is providing services of a much more substantial size, usually after some form of competitive bidding in a contest for award of a contract.

Independent contracting offers an extremely wide range of opportunities to become self-supportive and independent because a quite enormous variety of kinds of work can be and is being done under contract today. In fact, that is its major appeal as the basis for self-employment: Almost anyone can become a contractor. Some contracting ventures require substantial capital investment, but many, if not most, require little or no significant initial investment, which makes it a most appealing prospect.

The pages that follow will stress that diversity of opportunity in independent contracting, while presenting the business and entrepreneurial verities that are important in all business ventures.

Ironically, that very diversity and range of opportunities for independent contracting ventures poses a greatest difficulty in that there are so many fields in which one may contract and so many subjects to discuss as matters of importance in contracting, that all cannot be adequately covered in this book.

To handle that problem, I will not attempt to teach one how to do the work for which one contracts, but will focus instead on how to make a success of the *business* of contracting. I am assuming that the reader possesses whatever skills, abilities and competence he or she needs to do the work, but needs guidance in starting, building and managing a contracting enterprise, with special emphasis on the marketing aspects. Experience with other books and elsewhere in helping small businesses get started demonstrates that marketing is almost always the critical area, that one in which the independent entrepreneur needs help most urgently.

That will thus include major emphasis on marketing—winning contracts—but also guidance in how to make a start, setting up the business form, arranging a place of business, understanding contracts and negotiations and a number of other generalities about taxes, pricing services and other administrative tasks.

The Nature of Independent Contracting

Independent contracting is easy enough to define in strictly legal or technical terms, despite the opposition of the IRS in some cases. A practical, working definition is occasion for some extended discussion, for it is open to opinion as to what constitutes an independent contracting business.

Who and What Is a Contractor?

A contractor is someone who signs and operates under contract, right? A simple idea, if a literal definition is all that is required. But almost all business transactions represent agreements, even those inferred or assumed, rather than articulated in specific terms, and are thus contractual arrangements. A contract does not have to be written, or bound in a blue folder, with seals and signatures. A contract exists when one makes an offer, promising compensation (a *consideration,* in legal terms) of some kind, and another accepts the offer, if both parties are legally competent (of age and of sound mind) to enter into contract and what effort or activity is contracted for is a legal act or activity. Contracts and negotiations will be discussed in greater detail later in this book, but the main focus of this chapter is contracting as a business venture—as a way of doing business, or even as a way of starting an independent business venture.

Contracting has become something of a catchall term, identifying the many ways in which one may pursue any given trade or profession. In strictest technical terms, the minute you agree to do something for someone in exchange for some kind of compensation (a consideration, whether money or some other kind of compensation), you are a contractor. That agreement is a legal contract, even if it is not in writing. In other words, the contract is the agreement, not the piece of paper.

In practice, referring to contracts in the strictest sense as binding agreements, by far the majority of contracts are verbal and tacit; they exist by implication. When you sit in a dentist's chair, you understand that the dentist expects to be paid for dental services, and believes that you have that understanding and agreement without expressly stating it, although today you are likely to see a neatly lettered sign on the wall advising you that you are expected to pay for services when they are rendered.

Virtually every business transaction or trade exchange involves a contract; thus, anything you sell, service or product, makes you a contractor, technically. However, we don't think of a dentist as a contractor, despite the fact that the dentist enters into contracts with all his or her patients! We must therefore determine a practical definition of contracting for the purpose of studying independent contracting as practiced widely and representing a viable small business.

The IRS Has Its Own Opinion

The Internal Revenue Service (IRS) is not concerned with legal, technical or pragmatic considerations when the question of who is an independent contractor arises with regard to the kind of tax returns the individual is entitled to file. The IRS evidently perceives some advantage to them, in their tax-collecting duties, to regard the self-employed vendor of services as a temporary employee of the client, rather than as an independent contractor. Independent contractors are entitled to deduct certain kinds of expenses that an employee, temporary or otherwise would not be entitled to deduct, such as the cost of tools and unreimbursed travel.

This reluctance of the IRS to accept the claimed independent contractor status has been a difficult problem for certain kinds of independent contractors, especially those in technical fields such as computer and engineering services, where the client is likely to employ a number of such specialists on-site. Fortunately, at least for now, the IRS's opposition to independent contractor status does not greatly affect all such entrepreneurs, but is largely focused on those contractors who get most of their work through brokers and so really are subcontractors. The IRS finds reason and legal justification to challenge the claims of such subcontractors and to insist that they are temporary employees of the end-client—the broker's client, that is—rather than independent contractors.

Common and Uncommon Examples of Independent Contracting

For many, the word *contractor* summons up an immediate image of construction work, creating or renovating buildings, highways, bridges, paving roads and other such works. That is contracting of one sort, indeed, of the contracting we see going on about us constantly. In fact, almost any construction project, even the relatively small one, usually involves many contractors and subcontractors. The prime contractor for a construction job, who may be a company or an individual, undertakes overall responsibility for the project, becoming the prime contractor, and may do some of the work personally or only manage the project. However, hardly anyone does everything, even the contractor who handles much of the work personally. Almost every modern construction job involves a number of trades and crafts. The prime contractor, who normally needs the help of many others, therefore awards a number of subcontracts to plumbers, carpenters, electricians, tile-setters, excavators, masons, glaziers and sundry others, according to the needs of the project. Many of these subcontractors are themselves self-employed individuals who are specialists of one sort or another. As in almost all other fields of work today, building tradespeople have become more specialized. Where the carpenter once did everything that involved the use of wood and woodworking tools, today's carpenters tend to specialize in subcategories within their trade of carpentry. For example, today, the carpenters on a construction job often do not hang doors and door frames or install windows and patio doors. Instead, prime contractors tend to hire specialists who do such work. Such specialists tend to be more efficient in their work than the generalists, because they are experienced in and focus their attention solely on such specialized tasks. Usually, they have developed special methods and use special tools to speed the work, and have acquired experience that enables them to overcome typical obstacles and problems. They are thus often far more efficient at what they do than is the general carpenter, and fully justify their employment as independent contractors. Some specialists even base their service on their ownership of a special piece of equipment, such as a crane or earth-moving machine.

In more and more cases today, swarms of specialists are retained under contracts to do a great many of the tasks. Often, one specialist installs all windows and frames, another installs kitchen cabinets and others still do the wiring and hang the lighting fixtures, install glass tub and shower enclosures, lay down floor coverings and tile, hang draperies and handle all the other specialized work that goes into the modern home, apartment and hotel. Such specialists are often individuals and independent contractors, sometimes working as contractors to the client or end-user, but more often under subcontract to a builder who holds the prime contract for the entire project.

Retailers of all sizes, from the corner store selling draperies and blinds to the super-stores, have the same needs as prime contractors in construction and are themselves often prime contractors to the customers who buy home improvements from them. As the retailers, they and the smaller dealers need contractors to support their sales of items that require installation or maintenance. The dealer does not find it feasible to keep installers on salary because it is nearly impossible to keep the installers busy for 40 hours each week. Thus it makes much more sense to subcontract installation to independent contractors, and most dealers do just that. The dealer may arrange to send the installer or maintenance mechanic to the customer's home—the installer may even deliver the item for the retailer, in some cases earning an additional fee for the delivery service. Depending on arrangements the installer has with the retailer, the customer may pay the installer separately for the service, or the installation charges may be included in the sale price, and the dealer will pay the installation contractor.

An Independent Contractor Is Born

Independent contracting could be a perfect business opportunity for the individual who wants to be self-employed and is willing to do what is necessary to build a business: marketing; making arrangements with a sufficient number of prime contractors, dealers or end-users to handle their installations under contracts or subcontracts; and remaining busy enough to have a viable independent enterprise.

It really is not at all difficult to launch such a business. A large number of years ago I was employed for a time by a Miami manufacturer of aluminum-and-glass tub and shower enclosures to do the installation of his units. He sold these units primarily to contractors who were building tract homes and other residences, and I was paid an hourly wage to spend eight hours every day installing these units in new homes, apartments and hotels in Miami, Miami Beach and elsewhere in the area.

One day, while on the job installing my employer's units in a tract housing development, I was approached by the sales representative for a rival enclosure manufacturer. He told me that his firm did not employ installers, but contracted the installations out to individuals at a fixed fee for each installation. The firm and the sales representative were not happy with the contractor they were using. He asked whether I was interested in becoming an installation contractor for their firm.

I thought the proposition worth investigating and visited his employer that evening. After some negotiation, we struck a deal for a flat fee per unit. I then resigned my job, went out and bought a panel truck, with a small down payment of $10. (You may imagine how long ago that was!) For an investment of $10, I was quite suddenly in a business of my own as an independent contractor without losing a single day's work.

Before long, I began to win other clients, including dealers and department stores that sold such units, other local manufacturers and other sources, such as home improvement retailers whom I had persuaded to become dealers for the enclosures. In time, I also began to make special deals of my own directly with builders and acted as a sales broker for the units. In fact, I had salespeople for the units referring business to me, because then they could promise speedy and competent installation, a major factor in closing sales of any item that requires a specialist to install it. In other words, a reputation for reliability and good work produces word-of-mouth publicity and results in sales.

This is typical of how any independent contractor may begin in the various construction trades, some in general "fix-it" areas, troubleshooting and attending to a variety of typical small problems in home construction, but most as specialists of some kind.

Later, I was to become an independent contractor of quite a different kind of service: white-collar services, an area in which the number of independent contractors has grown enormously in recent years, in a general trend to what one committee in Congress had called the "mini-small business." It is almost as easy to become an independent contractor in these other fields as it is to become an independent contractor in the construction industries.

A Diverse Field

So contracting and subcontracting include not only construction and related blue-collar work, but also many white-collar industries and others that may be characterized as "gray-collar" industries. In all these fields specialization can make the independent worker more efficient in his or her specialty than would be the generalist within that career field.

The following list identifies only a few of the kinds of contract and subcontract work many individuals do today, operating from business locations in some cases, but more often from offices in their own, private homes:

- Writing scripts, newsletters, résumés and other items
- Editing and proofreading manuscripts for writers and publishers
- Delivering training programs
- Developing training systems
- Word processing
- Real estate appraisal
- Engineering support (all kinds of engineering)
- Office organization and design
- Market research (surveys, focus groups and so on)
- General research analysis
- Meeting planning and management
- Social event planning and managment (weddings, parties, etc.)
- Graphic arts or illustrations
- Computer programming
- Marketing management and services
- Interior decorating
- Consulting and brokerage
- Financial services

Consulting and brokering offer a numerous and diverse range of opportunities. By its nature, consulting is contracting, and so the independent consultant by definition almost is an independent contractor. Brokerage, also a rather diverse field, is inherently contracting. In fact, a product or service broker is a commission merchant and a contractor two times over in each transaction. As the liaison between the supplier and the buyer, the broker contracts with each, except in the case of the real estate agent, who usually contracts only with the seller of a property. The typical supplier of mailing lists, for example, is a broker, since he or she rents lists of customers, subscribers, inquirers, mail-order buyers and many others owned by others to end-users, and collects a commission for each rental.

Almost any kind of business may be brokered, including any services provided by the broker or others, although in some cases, brokerage is an impractical way to sell some services and products. Here are just a few common kinds of brokerages:

- Real estate sales and rentals
- Printing
- Temporary help
- Information
- Domestic help
- Staff recruitment/personnel services
- Discount and closeout/surplus sales
- Auctioneering (not all auction sales are brokerages, although the majority are)
- Commodity sales (in bulk)
- Financial services
- Finding

Gray-Collar Contracting

Even this does not suggest all the possibilities. There are services that are not exactly white collar, but still not blue collar (gray collar?), such as catering, facilities operation, messenger services, equipment maintenance and repair and handling dozens of other hands-on chores for clients on short-term or long-term contractual agreements. In today's economy, there is increasing need for help and services, some highly specialized, some not so specialized, that can be contracted for as projects for some defined term, to achieve some specified result, or engaged on short notice for an indefinite term and turned off as abruptly when the need ends.

The Larger Opportunity May Lie in Subcontracting

No company, not even the largest supercorporation, does everything. The larger the contract a corporation undertakes, the more numerous and diverse are the projects and subprojects required. Whatever the project, if it is a large one, the contractor needs supporting services of many kinds. Some of these support services are directly related to the main objective of the contract, while others are indirectly related to the main purpose of the contract, although not less important. The corporation winning a contract to build an aircraft, radar set or a new office building for a government agency (or anyone else) usually needs such direct support services as those already mentioned, but they also may need such indirect services as required to help them provide "documentation" to support the equipment. That will include such items as operations and maintenance manuals, parts lists, engineering drawings and whatever else the customer specifies as a requirement. The contract may also call for the production of training materials, such as manuals and audiovisual material, and even training in kind, actually conducting the training or training the client's trainers.

Why Prime Contractors Need Subcontractors

An organization undertaking a truly large prime contract almost never is able to do everything with its own resources: Few organizations find it practical to have the staff, physical facilities and other resources necessary to provide all its required tasks and functions. Instead, the organization becomes the prime contractor and issues dozens, sometimes even hundreds, of subcontracts to satisfy all those requirements it cannot do itself. The ability to find, subcontract for and oversee all the work necessary to get the job done is the responsibility of prime contracting. But whether the contract you undertake is a prime contract or a subcontract is simply a matter of whether you are contracting with the end-user or subcontracting to someone who is under contract to the end-user—that is, to the prime contractor. Keep in mind that you are an independent contractor to whomever the other party may be a subcontractor. In fact, in many situations it is much easier to be a subcontractor than a prime contractor.

Another Important Reason for Subcontracting

Aside from the fact that no organization does everything, organizations often have temporary overloads even in the things they do as their main activities or as in-house support functions. A firm needing help to handle a temporary overload of requirements may contract for temporary staff. But that is not always a practical approach, especially when the work is highly specialized. The better solution, especially when they need truly prompt reaction and sudden service (*quick response* or *quick reaction* services, in the jargon of the trade) often is to contract or subcontract the work to some established independent contractor who is experienced, equipped and ready to start work immediately. A great deal of contract work originates under such circumstances as the temporary overload or the difficulty of meeting an onerous delivery date. Many independent contracting ventures are based on providing overload services, and many independent contractors win their first clients by providing support of this kind.

Contracting Is a Special Working Arrangement

It is important to recognize and understand that contracting is not of itself a special kind of business or career; it simply is a way to practice a career specialty or conduct a business. If you are an engineer or writer, you don't stop being an engineer or writer by becoming an independent contractor; you merely opt for a certain way to practice and profit from your engineering or writing activities. When you write a book or article and sell it to a publisher, you do so under a sales contract. But when you write something for a client on a custom basis, such as a résumé or user's manual, you do so under a service contract. Both sales are made under contract, but in pragmatic, workday terms, it is only the second kind of writing that is considered to be independent contracting. That is because inherent in the idea of contracting is that the work, usually work of a special or custom nature, is done under contract.

That may be true even when the contract is to sell a product rather than a service, although the latter is by far the more common kind of contract activity. For example, if you have a warehouse full of sugar and you sell it to whomever wishes to buy sugar, that is a wholesale or retail sales business. In this example, you may be known as a dealer, wholesaler, retailer or distributor, but you usually would not be identified as a contractor. If, however, you contract to supply sugar and only after contracting to supply it set about buying the sugar, that is a contracting enterprise. And as a variant, if you agree to supply the sugar but make a deal with another organization to buy and ship the sugar with a commission to you, that is a brokerage business, although also a contracting venture, as all brokerage activity ordinarily is. In this variant example, you may be regarded as a commission merchant, rather than as a contractor.

As you can tell from this discussion, independent contracting is far from being a cut-and-dried proposition, nor is the ambiguity and difficulty of definition peculiar to contracting; it exists also with consulting and other business activities.

Finders

"Finding" is yet another kind of business activity that can be provided through independent contracting and subcontracting. It was listed earlier as a kind of brokerage because it is linked to brokering philosophically and bears some resemblance to brokerage, in that it bears the go-between or commission merchant stamp.

In some circumstances, those seeking to buy and those seeking to sell find it expedient to pay commissions or fees to anyone who can find an appropriate seller or buyer. It costs money to find supply sources, and it costs money to find buyers for what one wishes to sell; it may very well cost the buyer or seller no more to pay a fee to a finder than to do his or her own finding. During the energy crises of a few years ago, for example, owners of large stocks of coal offered finders' fees to anyone who could locate buyers, and sometimes buyers offered finders' fees to those who could locate relevant supplies. Sometimes someone seeking to buy something not easily or immediately available—such as a particular kind of real estate property or some special kind of business—will pay a fee to anyone who can find what the buyer wants.

This can be a business opportunity in connection with almost anything that is bought, sold or traded—even money. Finders' fees often are offered to those who can help find a certain kind of lender or investor. (This is a bit different than the financial brokerage listed, in that financial brokers make a full-time business acting as go-betweens in helping seekers of loans find lending institutions suitable for the borrower's needs, while the finder acts as a broker only temporarily and in special circumstances.) A finder may collect a fee from either party, buyer or seller, but not from both, according to which party wants the help of the finder and will agree to pay a fee.

Special Kinds of Contracting

One common situation in which a seller needs help is the volume retailer who is stuck with a large stock of slow-moving merchandise. The late Elmer Wheeler, often acclaimed as America's greatest salesperson, was sometimes called on—issued a contract—to help a merchant out in such circumstances. On one occasion a retail merchant, stuck with a large stock of long underwear, called on Wheeler for help. Wheeler solved the merchant's problem by making a huge display of the entire stock of long underwear and displaying a bold sign that said, "They Don't Itch!"

As a variant of that, the owner of surplus merchandise, the stock of a store that is bankrupt or out of business for some reason, or the executor of an estate that has a great many items to sell, will retain— contract with—an auctioneer to conduct a public sale of the inventory. That is another specialized field, and one that pays well. The auctioneer usually is paid a percentage of the all money received as a commission.

More commonly, a large department store will hold an annual clearance of slow-moving items in stock by holding a "warehouse sale," with boldly advertised discounts. However, the store may not handle its own sale, but may bring in an auctioneer to conduct the sale, under the reasonable premise that an experienced and skillful auctioneer will extract more money from the crowd than ordinary salespeople can.

Still Another Variant on the Theme

Contracting involves a special client-contractor relationship and the contractor usually has some kind of specialized skill. Still, contracting merely is the selling of a service, which also may or may not involve a product, just as one routinely would sell a product. A computer programmer works the same way when on a company payroll as when an independent contractor. That is true for other work done as an independent contractor. It is not the work that changes; it is the working arrangements, the relationship between the worker (you) and the boss (the customer), and also, usually, the amount of money exchanged.

It is, however, possible to launch a venture as an independent contractor providing a service or product other than anything you have done for a living, or providing a service that does not require any special skills. A former neighbor of mine and his wife provided a cleaning service to builders of tract houses, for example. As each house was completed, this couple went in with cleaning utensils and materials and scoured the house from top to bottom, preparing it for whomever were to be the buyers. An individual who is handy with tools may be able to easily pick up the skills and experience to install windows or hang doors, and begin offering services as an independent contractor for those chores. Or the part-time writer-hobbyist may seek contracts to write résumés, reports and other items needed in business.

"In Business" or "Self-Employed"?

As an independent contractor you are self-employed, but you are also in business. Practically, there is little difference in what you must do and be; you have the same tax obligations and business responsibilities regardless of how you regard your status. Psychologically, however, your mindset can make a substantial difference in how you conduct your business and perhaps even on the degree of success you enjoy. It certainly is likely to have a profound effect on your mode of operation and your plans for the future. If you regard yourself as being self-employed, there is a clear implication that you intend to remain an independent contractor, whereas if you think of yourself as being in business, there is at least an implication that you have the intention of growing into a larger business. Your self-view will have a significant impact on how you plan, the goals you set and your approaches to marketing, as well as the direction your business is likely to take. The "self-employed" mindset usually does not lead the contractor to excursions into related activities that the "in-business" mental orientation does.

When beginning to draft your business plan, you will have to face these and other factors. A business plan includes plans for the future, estimating where you are going, when you expect to get there and how you plan to get there. Because most plans are something less than perfect, they must be reviewed and adjusted periodically to meet the realities of the business world. You probably should review that plan at least twice a year for the first two or three years, and annually thereafter. Some individuals are successful without formal or organized planning, but the probabilities of success are undoubtedly much greater when you have a plan to guide you.

The Basics

Regardless of how you conceive and label your venture into independent contracting, whether as self-employment or in-business, you will be conducting a business and will need to satisfy certain minimal requirements of every business. I will list them briefly here, but we will discuss these in greater depth as we proceed.

- Financing: You may or may not need to raise a bit of startup/investment capital.

- Legal matters: There may be licenses, zoning and perhaps other legal matters you will encounter and have to deal with, depending on business decisions you make, local circumstances and perhaps on chance.
- Taxes: You will have income and FICA (Social Security) taxes for sure, and you may have other taxes, state and local.
- Insurance: It is essential to have insurance, at least health and hospitalization coverage and liability.
- Accounting: You must keep books and records; it's important to your management and the survival of your business, as well as to your tax obligations.
- Form of your business organization: Will it be a sole proprietorship, a partnership or a corporation? And if you don't opt for incorporation, will you be working under your personal name or doing business as (dba) under some trade name?
- Marketing: No business cannot exist without customers, and you can't find them without marketing. It is a prime essential of all business, more important, in its own way, than all the other matters combined.

Each of these items invokes other items and questions to be answered, and they are not unrelated to each other. The form of the business organization for which you opt, for example, will affect your legal and accounting requirements, and even your taxes and insurance. It will be necessary to discuss these, with the many alternatives and the pros and cons of the various alternatives. The answers are not easy to come by, in many cases, but require careful consideration and best estimates.

2

Start-Up: The Launch

The way to start is to start, but it gets just a bit more complex than that in this modern era: There are numerous business requirements. Still, there are other considerations, according to your own circumstances.

The Variables

There are many possible starting points and initial situations for the launch of any new independent or freelance contracting enterprise. You may start from a base of lengthy experience and many acquaintances in an industry and therefore with some substantial degree of self-confidence. You may start with an initial contract or two in hand or at least with one or two firm promises of contracts. Or you may start cold, expert in whatever trade or craft you intend to practice as an independent contractor with a firm determination to succeed, but with no other assets.

You may be without a job, and thus beginning with a relative carefree what-can-I-lose attitude, or with a sigh and a resigned it's-better-than-nothing attitude. You may be leaving a steady job voluntarily, in the hope and expectation of doing much better for yourself, or you may be leaving a job for which you have such a great distaste that any alternative seems preferable. In either of these cases, you probably have a substantial amount of gambler's blood. But then you may be one who is confident of your own ability to survive, in any case.

Finally, you may be in a good position financially and thus able to plan and prepare carefully and methodically, with no concerns about immediate income. But you may also have little but the last unemploy-ment-compensation check you received and thus have need to start earning from almost the first day. More than a few newly launched entrepreneurs are in that situation. (I know a man who started what would become a highly successful small business with his last $100 military mustering-out check.)

Major Steps

You will have to accomplish a number of tasks to get your new venture established. Your personal situation and your views of what is most important will almost surely affect your decisions as to what you must do first. Your major goals in starting out should include the following:

- Decide on a name for the business.
- Write a business plan.
- Arrange financing.
- Choose a kind of business organization.
- Set up a bookkeeping system.
- Decide how to handle accounting needs and tax reports.
- Obtain necessary insurance.
- Develop a marketing plan.
- Win the first contract.

The order of these steps, while reasonably logical, may not be optimal for all businesses. It may be necessary to do the last-named item first, and the financing step may not be needed at all; independent contracting often does not require extensive investment in either inventory or physical facilities, and can thus be launched without significant starting capital. Your principal starting capital may be your own skills and energy, sometimes referred to as "sweat equity." Aside from that, even if you must give priority to winning your first contract as a practical consideration, most of the items enumerated do not have to be in place on day one of your new enterprise; they can be your after-hours activities during the early weeks or even months of your new venture, so that you can spend the daytime hours producing income for your business. In fact, assuming that you do not need special financing, you probably need to give priority to the insurance concern, after attending to winning your first contract.

Insurance

Most businesses need personal and business casualty insurance. Actually, the distinction is arbitrary. What many regard as personal insurance for the self-employed person is really as much a business concern as it is a personal one. Still, for purposes of discussion, it is helpful to separate the two.

Personal Insurance

Aside from life insurance, you must have health and medical coverage. In these times, it is almost impossible and certainly unwise for the average individual to be without such coverage to handle medical costs for himself or herself, and his or her family. For many small, home-based business owners, this is a problem of considerable proportion. The majority of home-based business owners have been covered in a group insurance plan and have never had to face this problem before.

If you are already in a group of some sort through an employer, you may be able to convert your insurance to your name. However, the meaning of the word *group* is rather flexible in this context. You may be able to form or join groups offered by associations, families, clubs and other organizations.

Along with health coverage, you may wish to consider disability coverage, which often is included in health plans. This coverage can be especially important in a one-person enterprise, where your illness or other incapacitation may mean an abrupt and complete interruption of income.

There is also the matter of looking to the distant future and planning to provide for your retirement through some kind of retirement plan, which is also a kind of insurance. Most well-known are the IRA and Keogh plans, although any insurance broker would be glad to tell you about others.

Casualty Insurance

Casualty insurance is also advisable for most businesses. Most business counselors would consider it indispensable. It usually includes fire, theft and casualty coverage of your home and business physical facilities. But there is also the matter of liability. As a contractor, you may be held liable for losses incurred by a client if the losses are in some way connected with work you have done. Even if you are incorporated, it is wise to have casualty coverage of some sort in these litigious times, and not place complete reliance on the corporate shield. The protection afforded by the latter is not foolproof by any means.

Getting Help in Buying Insurance

Insurance is a mysterious, almost arcane, business to most of us, especially as it concerns casualty insurance. As others have in years past, I have been shocked and disappointed by what my insurance did and did not cover, learning the painful truths only when I had need of that protection I thought I had bought. When buying insurance, it is important to ask the following questions:

- Will losses be compensated at original values, at current values or at replacement values?
- Are premiums tax-deductible?
- How much will be deducted in compensating a loss?

If you do not know the answers to these questions when you buy insurance, you may be in for a shock when the day arrives that you have a loss and make a claim.

Insurance is a matter where most business people find it necessary to rely on expert advice, whether it comes from an insurance company person or an independent insurance broker who represents many insurance companies in acting as an insurance broker and adviser.

In my opinion, it is best to go with the latter option, to consult an insurance specialist, a consultant or insurance broker whose judgment you trust and whose integrity you trust. A broker can advise you expertly, as a rule, and is usually able to place your insurance coverage with any of many insurance companies, recommending companies whose rates and coverage are most favorable to your needs.

Writing a Business Plan

You can write a business plan at any time; it is best, however, to write it before you launch your business. When drafting your plan, do research to get as much data and ensure as much accuracy in the data as you can. You will have done a market analysis or some other sort of research to satisfy yourself that your idea is practical—that there is enough market for what you wish to offer to give you a fair chance of success. I have known even a major corporation to blunder in this regard. The corporation made a major investment in what the executives perceived to be a substantial niche market, but which subsequent experience proved to be too small a niche to be viable for more than an occasional sale.

Viability Research

If you are going to undertake independent contracting in a field in which you are experienced and are sure beyond doubt that there is ample business opportunity—i.e., enough prospective clients who will be willing to award contracts to independents such as yourself—you will not need to do the viability research. Otherwise, you must somehow verify that your idea is a practical one before you commit yourself. Investigate to make sure that there is a large enough market for independent contracting in the field of your choice to support your business. That may be step one of your business plan, but it probably ought to be taken prior to writing a business plan, perhaps the first thing to do if you are not absolutely sure that you are about to compete in a market where there is enough potential business to give you a fair chance for success. Only after having satisfied yourself that the market potential is adequate are you ready to proceed with the planning.

There are many books and computer programs to guide you in writing a business plan. However, there are business plans and then there are business plans. The business plan for a major business startup for which millions of dollars are needed as an initial investment, which will be markedly different from the business plan for your tiny startup of an independent contracting service for which little or no capital is required. The differences are neither superficial nor insignificant, and they are not merely a matter of degree; they are a matter of kind.

Objectives

Philosophically, a business plan should provide a road map for your venture. However, in a great many cases, perhaps the vast majority of cases, the business plan has as its main objective direct support for raising capital. In these cases, the very term *business plan* is almost synonymous with the terms *loan proposal, loan application* or *prospectus*. Its intended benefits as a guide to business success are real enough, but only in subordination to the primary objective of supporting the efforts to borrow money or attract investors for the start-up.

With the exception of that probably rare case where you need startup money to launch an independent contracting enterprise, the priorities here are quite different than those just described for a business plan written to support financing efforts: The primary objective of the business plan we refer to here is to accomplish the following:

- Think out and plan the venture in detail.
- Establish the mileposts by which to measure progress and success.
- Anticipate hazards and other problems.
- Enable you to detect hazards and problems promptly.
- Enable you to take corrective measures promptly.

The Marketing Plan

This is probably the most important element of your business plan— your strategy for winning clients and contracts. The most serious weakness a business can have is in the marketing area, resulting in a shortage of enough sales to support the business. The days of any business suffering this problem are surely numbered unless the problem is solved.

Briefly, your marketing plan should define clearly just what service it is that you propose to provide, your intended clients, and how you will reach them and make your offers and appeals.

Setting Goals

Your business plan ought to set forth the goals; the volume of sales that you expect to reach the first, second and third years, and whatever other mileposts you envision as marking progress and success of your enterprise. It also ought to include estimates of the costs of doing business, with details you can use for budgeting as a planned activity.

In short, you must *quantify* your goals, rather than merely qualifying them, and you must try to arrive at these estimates by some rational method, not by instinctual guesses. This compels you to face facts and fight the delusion that sometimes entraps the optimistic mind and can easily lure you into false hopes and unrealistic projections. A large part of the benefits of the business plan derives from having made realistic estimates and reviewing them frequently. In a short time, as you see how you are progressing toward those goals, you will know whether you are going to meet them, and you will decide what you must do in setting new goals: changing your marketing methods, cutting costs or otherwise making suitable adjustments. The point is not to see how near you can come to guessing what the future holds; it is to have targets and reference points by which to measure your progress and ascertain the realities to which you adjust.

The Business Organization

Other elements of lesser importance than your marketing plan and goal-setting, but still essential decisions and actions, are included in the business plan. One is deciding on the kind of organization your business will have. You have several possible choices, but two principal ones: If you are going into business with someone else, you can either go into a partnership, or you can incorporate. A partnership most definitely should be formed via a written contract between the partners, probably with each party having his or her own lawyer present. For the purposes of this discussion, let's assume that yours is to be a solo venture and as independent as it is possible to be. That means that your business organization will be either a sole proprietorship or a corporation.

As a sole proprietor, you are, indeed, self-employed. You pay your taxes, essentially the same as you would if you were an employee, although you will pay both ends (employee's and employer's) of FICA or Social Security taxes, as a self-employment tax, and whatever state and local taxes you are obliged to pay. Your tax reports are simple; they are personal tax returns. You are the business. It exists because you exist and do the work. You have total control, on the one hand, but you also have total liability.

As a sole proprietor doing business under your own name—e.g., Terry Jones—you need not register that name, except for any licenses your state or local government may require you to have. However, if you work under the name Instant Services or even Jones Contracting Service, you almost surely will be liable under some state or local statute dealing with fictitious names used for business. That is, you will be a "dba" (doing business as) entity and will be required to register that name so that anyone can determine what individual is doing business as "Instant Services."

It is relatively easy to incorporate. In many states, the proprietor can even hold all the offices on the corporation's board. In many states, incorporation is as simple as filling out a one-page form and paying fees of $50-$75, and you can buy a seal, bylaws and all necessary forms quite easily for a few more dollars. Corporations are created under state laws, which vary considerably, so it is advisable to research your state's incorporation laws. It also may be a good idea to consult a lawyer when incorporating. The final decision is yours, of course, but a lawyer can advise you on the pros and cons of incorporation and of each of the types of incorporation available to you.

Incorporation has certain advantages and disadvantages. A corporation is a legal entity. The corporation becomes your employer and must pay certain extra taxes and fees, depending on the state in which the business is incorporated and the state in which it is headquartered. While it is not necessary to be incorporated in the same state in which you do business, it probably is advisable to do so.

The tax advantages of incorporation are not as significant as they once were, compared with taxes a sole proprietor pays, although there are some other advantages. The corporation affords you some protection from liability, for one, but again, this protection is not all-inclusive or foolproof; you still should carry liability insurance. There is no need to register a corporation as a dba, however you name it, since you will already have been registered as a corporation. In fact, you will have to wait to receive your incorporation papers while the state makes sure that the name you chose is not already in use by an existing corporation.

Accounting

The accounting function is only slightly less critical and important than the marketing function. Its importance is due to a variety of factors.

First, the Internal Revenue Service requires every business to keep records. Most people interpret that as a requirement to keep account books, although that is not so. If you keep all your records in a box, as self-employed individuals of yore did, you have complied with the law. It is increasingly more practical, however, to keep your "books" in a computer financial management software package.

The second reason to keep records is as a guide to successful business management. Accounting is principally a management tool. It tells you what and how you are doing. It allows you to compare your liabilities and assets (dollars spent versus dollars received). It guides you to the trouble spots and provides the information you need to ascertain, analyze and overcome them. Used properly, it helps you find and stanch the business wounds before they become fatal. On the other hand, it also helps you uncover the opportunities to increase your business success.

Who Will Do Your Accounting and Tax Returns?

You have at least three options to handle your accounting and tax-reporting needs:

1. You can do it yourself.
2. You can have an accountant do it for you.
3. You can use a combination of these methods.

Doing it yourself is not difficult if you operate as a sole proprietor. Your tax reporting is the simple individual tax return of the self-employed individual. The books you need are very simple. You can keep records manually in a single ledger or even a notebook or buy a patent system designed for the small business; office-supply stores carry several such systems. You also can buy a software program and keep your books on your computer. You can do these things whether you are a sole proprietor or a corporation, although the books and taxes become considerably more complex when you are incorporated.

The second option, turning your accounting function over to an accountant, who will do it all for you, including your quarterly and annual tax reports, is convenient and expensive.

I favor the third option, dividing duties by handling the bookkeeping yourself, and having an accountant prepare your tax returns. By keeping the books yourself you not only save the cost of having a stranger keep your books, you are more likely to stay abreast of how your business is doing and alert to any exceptional condition. Without careful and continuous monitoring, businesses can lose money for a long time before the owner becomes conscious of it. By the time the owner is aware of the problem, the business may be in such serious straits that it is difficult to mount a rescue effort. This type of situation has resulted in more than one business bankruptcy.

Bookkeeping is a simple process for an uncomplicated, one-person independent contracting business. But regardless of whether you keep your books yourself, you ought to have a basic knowledge of what accounting is for a business such as yours.

What Are the Basics of an Accounting System?

In its simplest terms, accounting is about money. It's about how much money you take in, how much you pay out and how much if any, you get to keep. Obviously, the objective is to take in more income than you pay out as expenses or costs and keep the difference as your profit.

Business does need to make a profit. Many inexperienced young business owners make the mistake of counting their own salaries or draws as profit. It is not. Your salary is part of your business costs; only the money remaining in your hands after you have paid your own salary and all other costs of doing business is profit. It should be banked as a reserve. You will need it for replacement of equipment, maintenance, expansion and other situations that are likely to arise.

You therefore need to keep an account of what you spend and what you receive, and the balance between the two. You need to keep records in such a way that you can identify each expense item and income source. A large accounting system might contain various ledgers—literally books—but you can do the whole thing with a single book.

A simple bookkeeping system is illustrated in Figure 2.1. Records of receipts (income) and expenses are kept on a month-by-month or week-by-week basis. You can set up as many account numbers as you wish for each category. The left-hand column presents figures carried over from the previous month, the middle column is this month's figures and the right-hand column is the total for the year to date. The last column will become the left column on the next month's page. Thus, you know, almost at a glance, how much you have taken in and how much you have spent to date, as of the end of last month.

You need one other page: the day journal or diary. The purpose of the day journal is to log income and expenses as they occur during the month. At the end of the month, you total the figures for each account number and post the totals.

Many other elements are included in an accounting system, such as depreciation of certain items as they age, amortization investments, striking profit and loss statements and balance sheets and other records. It is helpful, but not really essential, for the one-person business to keep these kinds of records. Capital item depreciation is, however, important to understand. A truly large item, such as a truck you buy and use for business only, is an expense, but the IRS will not permit you to "expense" this, or charge it off to your business completely in the year you buy it. Instead, you must spread the expense over the life of the vehicle, charging off a percentage each year. You can, however, expense some of the smaller capital items, such as computers. Many of these things will become clear to you in the instructions that accompany small business accounting systems, whether they are paper- or computer-based systems.

FIGURE 2.1 A Simple Bookkeeping System

Month _____ 19 ____

Expenses	Total last month	Total this month	Total to date
8010 Accounting _____	_____	_____	_____
8020 Rent _____	_____	_____	_____
8030 Automobile _____	_____	_____	_____
8040 Books/subscriptions ____	_____	_____	_____
8050 Insurance _____	_____	_____	_____
8060 Postage _____	_____	_____	_____
8060 FICA _____	_____	_____	_____
8070 Withholding tax _____	_____	_____	_____
8080 Office supplies _____	_____	_____	_____
8090 Dues _____	_____	_____	_____
8100 Repairs _____	_____	_____	_____
Totals: _____	_____	_____	_____

Month _____ 19 ____

Receipts	Total last month	Total this month	Total to date
6010 Consulting _____	_____	_____	_____
6020 Royalties _____	_____	_____	_____
6030 Mail order _____	_____	_____	_____
6040 Interest _____	_____	_____	_____
6050 Miscellaneous _____	_____	_____	_____
Totals: _____	_____	_____	_____

Choosing a Business Name

A government executive, who received many visitors seeking business, once observed, "the bigger the name, the smaller the company." He was amused by such names as "Federal Capital Data Systems Information Corporation," when he knew that behind this grand name was a company boasting only one or two employees.

I incorporated my own business some years ago as HRH Communications, Inc. I wish now I had used an even shorter, simpler name, although I deliberately chose a name broad enough to justify a wide variety of business initiatives under the corporate umbrella. Incorporated only because I had a special reason to, I would have been perfectly content to conduct my business under my own name as a sole proprietor and, in fact, do so conduct some of my activities.

There is no glory and little, if any, general benefit in having a grand name of many syllables. We live in a rather sophisticated world today—radio and television have penetrated to the most remote corners—and the public is not as easily impressed as they once were. Even in earlier times, many companies were launched under the owner's name and managed to grow into large corporations. Milton Hershey used his own name to start his chocolate candy company. Gail Borden launched his canned milk company under his own name, as did Alfred Fuller, of Fuller Brush Company and W.A. Sheaffer, of the Sheaffer Pen Company. I find something wholesome and reassuring in a simple, direct name, especially when it is the proprietor's own name, and includes a clear idea of what the proprietor does for customers.

If you want to use your full name in your company, for example, Jane J. Jones, you may make good use of a blurb of some sort to amplify your basic message. You might say on your cards and brochures, "Jane J. Jones, Graphic Arts Services."

Above all, I exhort you, don't get clever in designing a business name. Semantic cleverness is a deadly disease in marketing, and your name should be an asset to your marketing: It ought to be straightforward and clear. Subtlety has no place here. If your prospects cannot make out what your name means, they simply will go on their way without another thought. If you are a computer consultant and you call yourself "Ones and Zeros" or "John J. Jones, Ones and Zeros," that is clever enough, but who knows what it means? Even the computer-literate prospect is not likely to suspect that you are cleverly telling all that you contract to write computer programs.

Logos, Stationery and Other Business Appurtenances

At the risk of making myself unpopular, especially with designers and printers, I will share my honest opinion of business "image" materials. I am mystified by the insistence of some newcomers to their own tiny business ventures on purchasing expensive and elaborate business trappings such as cards, letterhead and logos. Some who are not yet turning a profit, much less earning a living from their new enterprises, cheerfully invest many hours of their time and hundreds of dollars to have a designer create an artistic logo for them. And all too often, the design that cost many dollars is something no more complex than a stylized initial or monogram. They then spend still more hours and hundreds of dollars more on fancy, engraved stationery in complete sets—matching business cards, letterhead and envelopes.

Of course you should have a professional-looking business card and presentable letterhead and envelopes. A logo? Perhaps when you are as big as General Electric or IBM you can afford that luxury, but that time is a long way off, and no one expects you to have a logo or cares about it. It won't buy you anything, even if it is clever, eye-catching and arouses favorable comment from everyone who sees it. What you are most in need of, at first and for some time after, is business—contracts and income—not praise. Many who so indulge themselves justify the investment by insisting that those who see their logos and stationery will be impressed and remember them. Maybe. And maybe being remembered will produce an additional contract or two, eventually. But you need contracts now, not eventually. "Eventually" doesn't pay the bills. You need to invest your marketing dollars in whatever activities will be most effective in producing those contracts now and positively, not in some undefined future and maybe.

You will need some kind of general brochure, and you probably will need a capability brochure. These are among the marketing considerations which we will discuss later in this book.

Financing

While few independent contractors need significant startup capital, the possibility does exist that you may need to make a significant up-front investment for inventory, tools, fixtures or other physical facilities and resources.

For a great many entrepreneurs, the main objective of a business plan is to help in raising capital. Why that is so should become clear with even a brief discussion of the subject of financing.

Debt Financing

There are two kinds of financing: debt and equity. In its simplest terms, debt financing is borrowing. You may borrow from friends and relatives or from lending institutions.

One example of debt financing, a mortgage, is known as a secured loan. The property is collateral for the loan. As long as the bank sees the market value of the house as larger than the amount of the loan, the loan usually is granted. The property is its own collateral and the lender is protected by holding a title, lien or other surety.

If you wish to borrow money to invest in a business venture and you have collateral to protect the lender—a property in which your equity is great enough, securities of some sort or other assets that can be liquidated, if need be—you will probably have little difficulty in getting a loan.

If you have no collateral or you wish to borrow considerably more than you can back up in collateral, you have an entirely different situation. Banks and other lenders are usually reluctant to undertake this kind of loan—an unsecured loan—because they perceive the risks to be rather great. This is what makes the business plan so important to the individual who wants to borrow money strictly on his or her business prospects. It takes a great deal of persuasion to get a loan under those circumstances.

Equity Financing

Selling equity—a portion of your ownership—in your business is an alternate method of financing it. There are a number of ways to do this. You can take in others in limited partnerships, incorporate as a public corporation and sell shares and seek venture capital from professional venture capitalists, among other methods. You can arrange to confine investors' involvement to the earning of dividends and appreciation of their shares, so that they do not have a voice in management of the business. However you arrange matters, you have always the basic issue of investor confidence in your business. Inspiring or gaining that confidence in your proposed venture is the principal objective of the business plan if you write it to help find substantial capital to finance your business. In that respect, the business plan to support debt financing is not different than the business plan to support equity financing.

Taxes and Deductions

You will have to pay a number of taxes to federal, state and local governments. FICA or Social Security, also known as self-employment tax, will be one of your most prominent taxes. The bulk of the taxes you will have to pay to the IRS will be that self-employment tax and income tax, both taxes on your earnings.

You can't do anything about the tax rates, of course, but you can do something about the gross amounts taxed. You can and should take all the deductions to which you are legally entitled. It is against the law to *evade* taxes—fail to pay the taxes you owe—but it is not against the law to *avoid* taxes by taking full advantage of every legal loophole allowed.

The Home Office

The largest deduction for most home-based entrepreneurs and probably for you is the deduction for your office, shop or other working facilities in your own home. If you can manage to work from your own home without an establishment elsewhere, you will enjoy a great advantage. You will have a substantial business deduction for rent or mortgage payments and portions of all facilities—telephone, electricity, water and others—that your business uses in your home. That, in turn, will reduce your personal living expenses. But there are certain rules governing deductions for the home office.

Dedicated for Business Use The first rule is that any space you use at home for your business must be dedicated to that use, and not used for anything else. You can't take a deduction for using the dining room table once in a while to do some business chore. In my own case, two of our bedrooms are set aside as offices and are used only as offices. Their cost to us is therefore deductible as a business expense. Conceivably, our dining room could be a deductible office, if we chose to make it one and used it for that purpose only. On the other hand, it is difficult to make a case for the use of a portion of one's bedroom as an office meeting the requirements. Conceivably, however, you could cordon it off with some kind of room divider and thus qualify it as dedicated to business use only.

Since we live in a high-rise apartment building in which our rent covers all utilities—water, gas and electricity—we take a proportionate share of the rent overall as our office deduction, as that automatically covers a fair prorated share of utility expenses. We also have two telephone lines. One is entirely deductible as a business expense because it is used only and entirely for our fax and computer modem connections; the other is a voice line, used for business and personal matters, and so deductible only insofar as it is used for business calls. We make a prorated deduction that the IRS so far has accepted without challenge. We also keep a post office box for our business mail, so that cost is deductible as well.

Principal Place of Business The second rule is that the home office must be the taxpayer's principal place of business. That was the principle upon which the IRS was upheld by the U.S. Supreme Court when they disallowed an anesthesiologist's claim of a deduction for his home office. The Court held that the anesthesiologist's principal place of business was in the hospital, where he administered anesthesia, and that he was thus not entitled to take a deduction for an office he maintained in his home to keep his records and from which to issue his invoices. That opinion made it clear that the burden is on the taxpayer to demonstrate that his or her home office is his or her principal place of business.

The problem lies in defining principal with respect to one's office. Among the tests are these:

- Does the taxpayer see clients there?
- Does the taxpayer perform the service or produce the product of his or her business there?
- Does the taxpayer spend most or a great deal of his or her time there?

This is not to say that the answer must be affirmative in each case for the home office space to qualify as a deduction, but each affirmative strengthens the taxpayer's case.

The 1706 Problem

The IRS has little affection for the idea of independent contracting and would prefer that all independent contractors become their clients' temporary employees, with all the attendant withholding of taxes, fringe benefits and paucity of individual tax deductions on their annual returns. To that end, the IRS has tried diligently to disallow independent contractors' claims that their independent business ventures should be given the same tax considerations as true of every other business in the United States.

Unfortunately, the Tax Reform Act of 1986 includes the now infamous Section 1706. This exposes you to the IRS' 20 questions that presumably define whether the taxpayer is an independent contractor or a temporary employee masquerading as an independent contractor. Following is a summary of these questions to give you an idea of their general orientation:

- Who controls your daily work? Is it you or is it the client giving you orders each day?
- Do you have many clients during the year, or is all work for only one or two clients?
- Where do you do the work—on the client's premises or your own?
- Who sets the hours of work, you or the client?
- Can you and do you work for more than one client at a time? (Do you have concurrent contracts?)
- Are you paid by the hour or by the job?
- Do you provide the tools for the work?
- Are you responsible for all necessary training?
- Do you undergo the risk of losing money?
- Are you immune to being fired arbitrarily?

These questions raise the issue of independence. The effect has been not merely to disallow the tax returns submitted by many independent contractors, but also to alarm many clients and discourage them from awarding contracts to freelance operators.

The chief effect has been seen primarily in cases involving a third-party contractor. Contract brokers used extensively in the computer field, supply a given number of specialists to work on a client's premises. This firm, the prime contractor, brokers the work by awarding a number of subcontracts to individual independent consultants and contractors. The IRS insists that the latter are temporary employees, not entitled to file as independent businesspeople receiving 1099 forms from their principal client, who is, of course, the prime contractor. The prime contractors or brokers, in turn, are most apprehensive that the IRS will demand that they withhold taxes and pay fringe benefits for all these independent contractors—subcontractors, technically—and so are less and less inclined to accept the independents as such, believing themselves to be forced by the IRS to be "job shops," employers of technical or professional temporary employees.

For the moment, at least, the problem appears to be centered on the third-party situations that involve a great many subcontractors. The truly independent contractor who negotiates independent contracts directly with clients, rather than with brokers or prime contractors, probably has no problem other than the possible reservations some clients may have about negotiating such contracts. Every independent contractor should heed the problem and do everything possible to avoid any resemblance he or she may have to the temporary employee.

3

An Introduction to Marketing

Marketing is a simple concept. In practice, however, it becomes considerably more complex.

We All Market

Everyone markets. No one and no organization exists without marketing. Marketing extends and perpetuates one's existence, whether that of an individual or an organization. In fact, that is one workable definition of marketing: an activity to perpetuate one's existence. What we tend to think of as marketing, the process of selling goods and services for money, really is only one kind or application of marketing. Military organizations market when they recruit enlistees; associations market when they have membership drives; charities and politicians are marketing when they have fund raisers, soliciting contributions; and politicians market when they solicit your vote. We even market on a personal level when trying to get a date for the prom or sending out résumés and letters applying for a job or for college admission. You also will have to market to launch and sustain your independent contracting venture.

People Are Your Market

Despite the wealth of literature attempting to explain its meaning, there is nothing mysterious about the word *market*. A market exists for a product or service when there are people who will buy it. Sometimes markets are defined geographically, such as the New York market or the midwest market, again, however, that refers to the customers living there, not the region.

There is the teenage market, made up of children and young adults 13 to 19 years of age, who have quite a bit of disposable income depending on their parents' incomes. On the other end of the scale is the seniors market, made up of people past 60 and, to a large degree, retired and living on pensions and savings, often relying also on part-time jobs.

Obviously, these are quite different markets. While some goods and services are viable in both markets, others sell well in one but not in the other. One of the important functions of marketing is to properly match items to the people who will purchase them.

Markets vary according to what you offer. Seniors certainly are not a market for military recruiters, but they could be a market for cruises, second-career plans and retirement condos. They also may be excellent candidates for do-it-yourself offerings, since most have considerable time on their hands, and many are interested in second careers of one sort or another.

All markets are segmented. Take the market of seniors/retirees. Within it you will find some of affluence, who are ready for a life of leisure and pleasure. They will embrace the cruises, the trips and the other recreations. Others will buy recreational vehicles and take to the road. Small segments constitute niche markets. These are small, special groups of people, special in that for some reason, they are more likely to buy a given product or service than is the general public or the main market for the item. A small portion of retirees will buy house trailers.

To switch to the teenage market, keep in mind the difference between a 13-year-old and a 19-year-old. The former might be a good match for gym shoes or candy, while the latter might be looking for help in drafting a résumé for a first job, or in finding scholarship funds for a college education.

It is that smallness that makes these niche markets, and that is also what makes them an attractive target for the independent operator. Larger competitors may have overlooked the niche or they may be aware of it but deem it not worth mounting a marketing effort to capture it. That may be true enough and usually is, but it does not mean that it is not worth your efforts to sell to it. More than a few small businesses operate profitably and prosper in niche markets.

Marketing or Selling

Most of us tend to use the words *marketing* and *selling* interchangeably. Actually, they are not the same, although no great harm is done by confusing the two terms. The purpose of marketing is to sell whatever it is that you propose to provide to those who will have use for your service or product. Selling, to be technical about it, is the final act in the process of marketing—finding customers for your product or service. Marketing takes in all the activities preliminary to and leading up to the sales effort, including determining what to sell, how to price it, how to advertise it or present it to prospective customers, identifying the best prospects, determining how to reach them and setting the stage properly for a successful sales effort. The so-called "sales strategy" really is a marketing strategy.

The first stage of marketing, deciding what you will sell, is perhaps also constrained by determining exactly what you *can* sell. Your options usually are limited by your own knowledge and skills, the nature of the markets within your reach and perhaps a few other conditions that are not completely under your control and perhaps not even to a small degree; there is always the matter of being absolutely sure that you can deliver the product or service you promise, as you promise it.

To be an independent contractor means, usually, providing some kind of service, although products may be involved in some way. You can offer a general set of services within some field—training systems development and support, for example—or you can specialize and confine yourself to one or two specific kinds of support, such as training systems in a limited number of subject areas or in a limited number of media. That will depend on your preferences or your capabilities. Seminar producers and sponsors, for example, usually will require that their contractors provide the material and the delivery, but some have "canned" seminars and hire professional speakers to deliver their material from the speaker's platform.

Market Research

The first major goal of market research is to assess the market potential to determine overall viability. That established, the research must then go on to answer such questions as these:

- What and where are the markets?
- How large are they?

- How ought they be prioritized?
- What kind of appeal is most likely to win?
- How can you reach them with your sales appeals?

Most of us, as small independents, are not in a position to contract for or conduct our own formal market studies. We have to operate largely on our own estimates, and even these offer little more than the best guesses we can make.

As a first step, you can list all the possible markets you believe to be good possibilities. That list will be influenced by what you offer. If you are confined by your capabilities—perhaps word processing, light copy editing and proofreading are all you believe you can offer—your choices will be limited to those services. You will not be able to pursue those prospective customers who want writing services, unless you wish to subcontract or broker related services you cannot or do not wish to do. That doesn't mean that you do not have options with regard to the markets you will pursue. There is a wide enough demand for word processing, light editing and proofreading to represent a broad array of possible markets for your service, regardless of whether you choose to provide related services. For example, when working as a writer handling projects for clients, I often had to provide the illustrations; I thus had to subcontract this part of the work if I wanted the contract.

Once you have decided what you will offer, you can begin to identify all the possible markets as the first step in your market research. Let us identify a few of the markets in general terms for such services as we have established here in our hypothetical case:

- Individuals
- Local merchants
- Companies
- Government offices
- Associations

You must translate this general list into a more closely defined list. You can do this by thinking out the subcategories, determining which of these subcategories are of significance for you and what kinds of word processing needs each of those subcategories might have.

Individuals

One subcategory that comes to mind immediately is students. What is your situation with regard to students? Is there a university or two in the area? Your primary student niche would be college students, and so it is important to assess the possibility from the viewpoint of availability of a student market.

Many students of high school and college age will have their own computers and word processors today; still, many might want professional help to produce their term papers, theses, résumés and whatever else they may need in the form of text on paper. They may not have the most modern of printers, for example, and might want something better than a dot-matrix printer can produce. Or they simply may be unable to turn out an interesting and professional-looking presentation and be quite willing to pay for your services to produce a more attractive product. Thus, you may define your markets in another reference: the service or item involved in and reason for the required service.

That brings up the consideration of other individuals, ordinary citizens, who may be good prospects for you. Anyone may need a résumé organized and printed out, or a letter put into a professional-looking presentation. Some local professional or executive may need a speech or a paper composed and printed out. Thus the professionals in your area—physicians, dentists, lawyers, architects, engineers and many others who belong to professional societies and career associations—are a market, or perhaps a niche market if they represent a segment of a larger market in your service area.

Local Merchants

Local merchants are likely to have a variety of needs that your services should be able to satisfy. That could include preparing sales letters, writing collection letters, composing brochures, preparing fliers and other tasks. That could vary quite a bit, depending on the nature of the merchant's business. Consider just a few of the possibilities:

- A furniture store would probably send out sales letters and sales announcements much more often than would a drug store.
- A restaurant might want a daily lunch menu or a list of dinner specials, and might want them laser-printed because they would probably not need more than a couple dozen copies each day.

- Local print shop owners might have a great many needs for your services if they offer composition services to their customers, and may even want to subcontract very short run printing to you to be laser-printed.
- Any merchant may use a weekly or monthly bulletin of some sort to send out as announcements.

Almost any merchant could make use of your services to do any of these things, as appropriate to their businesses, and they probably should. It may be that they don't simply because of inertia; it requires initiative to even conceive of it, let alone arrange to do it. For that reason, you may wish to consider sending out a letter suggesting these things and inspiring local merchants to adopt some of these ideas, rather than wait on the remote possibility that the inspiration will strike the merchants. But more on that subject later.

Companies

Larger business organizations, companies and corporations in your area have the same kinds of needs, although on a much larger scale. They tend to have substantial marketing campaigns, often large direct-mail programs. They publish newsletters and catalogs; prepare press releases; and send out many kinds of materials that they first must generate.

Many of these organizations have ample in-house facilities for their word processing and laser printing needs, yet they still may be excellent prospects. In fact, the organization that has its own in-house facilities often is the best prospect. They have the in-house facilities because they do a great deal of such work; and because they do a great deal of such work, they often are overloaded and in need of extra help on a short-term basis, one that is best satisfied by contracting with someone local. I have found the organization with its own capabilities to be an excellent source of business, especially when the problem is one of pressing schedules. In such cases, organizations usually will pay premium prices to induce you to work evenings and weekends, and they will be grateful to you, regardless of what it costs them. You are a rescue service, and they are likely to remember you the next time they need to be rescued—or maybe even when they are not in big trouble, but can use some help.

Government Offices

Government offices are especially subject to peaks and valleys of activity, and are often temporarily overloaded and quite eager to contract out their overload. Later, we'll discuss how to pursue government contracts at all levels: federal, state and local. They are excellent markets for many reasons.

Associations

Associations are everywhere. They include fraternal or social organizations, such as the Elks or the local home and garden club; business or trade associations, such as the Lions and Rotary clubs or the Processed Foods Manufacturers Association; career- or occupation-oriented individuals, such as the local writers' club; and many other bases, such as veterans' groups and labor unions. Some, like the Elks, are national and even international, while others are strictly local, such as the local writers' club. All have needs, and even the local groups can be large enough to award contracts for whatever needs they have.

Other Nonprofit Organizations

Associations, governments and unions are not the only kinds of nonprofit organizations. Charities, such as the Salvation Army, certification organizations and other groups are other varieties, and many of them can be business prospects also.

Needs Are Universal

Many needs are not peculiar to any single market. That brings up another goal of market research: Identifying markets by their needs, rather than by who or what the clients are. You can start by ascertaining the needs of prospective customers and comparing them to what you offer, or you can go the other way around and reason from needs to client identities. If you use needs as the identifier, other classifications tend to fade in significance. That is, almost anyone in any of those general groups of individuals and organizations listed earlier might have need of a résumé, a paper or a letter, so the need is the only significant factor. In that case, rather than attempting to analyze who can you reach or how can you classify people as prospective customers, try to visualize the many uses to which your skill can be put and what needs it can accommodate or problems it can solve. Then you can begin to cast about in order to estimate who may have such needs.

This approach gets you to the same place: matching the market (people you can reach) with the need you can satisfy. Which approach works better? That varies, from one case to another. It's one reason that marketing is far more art than science.

Marketing Is Consulting

It has long been recognized that selling is consulting. Consulting is problem solving, and so is selling and, by extension, marketing. Customers buy what you sell to solve their problems or satisfy their needs. To cite an old saying, the key to business success is to "find a need and fill it." That is easier said than done. We all have needs, lots of needs and all kinds of needs. The challenge is not in "finding a need" or to "filling it"; millions of people do that every day, many of them for minimum wage. There is something more to it, of course. The saying's originator probably meant that you should find a want that no one is filling very well, and fill it much more satisfactorily. Keep this idea in mind as we proceed to talk about wants and needs.

Wants and Needs

To understand how to market effectively, you must visualize two kinds of needs: felt needs and created needs. The difference is probably most easily explained by an example: The customer who enters a shoe store to buy a pair of shoes is responding to a felt need. He or she *feels* the need for new shoes, and all that remains for the salesperson is to help the customer satisfy that need. Most sales clerks will do that. However, let us consider the sales clerk who is a *salesperson*. The alert salesperson sells the customer shoes, but also sells boots and socks and perhaps whatever other items the store has to sell. That salesperson has created needs by making the customer decide that he or she also needed the boots and socks, needs he or she did not feel before entering the store.

The bulk of the marketing/selling problem—and in the marketing/selling opportunity—lies in the ability to create needs, to point out to the customer the need not yet felt and persuading him or her to feel it and respond to it. Response is, of course, in some proportion to the degree of your success in making the prospect feel the need. Marketing success—business success itself, in fact—is directly proportionate to success in creating needs, in the sense used here to awaken in prospects consciousness of a want or need for what you have to sell.

Passive or Aggressive Marketing

Consider the example of "Joe the Barber." No matter how good a barber Joe is, a customer will not patronize his business until the customer feels a need for a haircut. Joe's marketing strategy is to create a good professional image and make himself known, so that his customers single him out as the supplier when he or she feels the need for relevant services.

That argument is good to a point, but it falls short when it comes to the point of waiting for the prospective client to feel the need. Waiting for the prospect to decide to become a client is passive marketing. Waiting can be lengthy and ultimately fruitless. Joe has to find ways to induce customers to come to him, even when they do not think they need a haircut. Perhaps they need a trim? A facial massage? A scalp treatment? A beard trimmed? Perhaps they could take advantage of a special deal?

Successful marketing gives prospects reasons to feel a need. That may be brought about by fear marketing—pointing out what dire consequences may result if the prospect ignores or neglects the prospect of future needs, or gain marketing—pointing out what the prospective client stands to gain by acting now, before the need is acute.

The latter example is aggressive or creative marketing. The very word *marketing* implies an active, or aggressive, effort to find and sell clients, rather than waiting for them to come to you. Another example of aggressive marketing is seen in one example of a carpet merchant who runs a continuous sale, with heavy advertising, for years. The fact that the sale runs on week after week and month after month for years means it isn't really a sale at all. Those prices are in force at any and all times. But the public may not care about philosophical distinctions; they take advantage of the endless "sale" every day.

Creating Needs

By far the best way to create a need is to offer something new, different and of obvious utility in satisfying a need better than it has ever been satisfied before. Anything that meets these criteria is all but assured of near-instant success.

The automobile is one outstanding example of this. It satisfied the need for transportation better than anything that had gone before, including bicycles and horses, for local and short journeys, and trains for longer journeys. The public rapidly embraced the automobile. It was fair to say that a new need was created: a need for automobiles. Of course, the need had not existed before because the automobile had not existed.

Motion pictures are another example. Motion pictures were an improved means for satisfying the need for entertainment. Vaudeville survived for many years, and live theater survives to this day, though it does not enjoy the popularity it once did.

Radio was a great success, and television an even greater success, because they provided entertainment in the home. Together, they sounded the death knell for vaudeville and even night clubs. More recently, the advent of video recorders and laser disks have made the movie palace an endangered species. Each of these inventions filled a need better than its predecessors for home entertainment, and both found swift success. Remember that before the advent of radio, home entertainment consisted of kaleidoscopes and card and checker games, and not a great deal else.

Of course, most of us cannot count on inventing the next revolutionary new inventions. Still, the philosophy is unchanged: Try to find better ways than any extant for solving problems and satisfying needs. When I became a proposal consultant, it was still a relatively new idea, and I could identify few direct competitors. But a service or product does not have to be new; if it is significantly better than anything that existed before, that alone justifies it and stirs a demand—a felt need—for it.

There is still one catch in all this: while most of us believe that what we offer is superior to all others, that is only our opinion. For marketing success, prospective buyers must recognize the product or service as a better way, which leads to another important aspect of marketing, that of influencing prospects' views and attitudes toward what you are offering.

Educating the Consumer

As has been said and demonstrated to be true often enough, you cannot sell a customer anything he or she truly does not want. No, that does not contradict what you just read about creating needs. It affirms it. Creating needs means persuading customers to feel the needs, to want what you are offering. It does not become a felt need until the customer is completely conscious of it as a felt need. But even that misses the true point. In a greater sense, all true needs are felt needs; you cannot sell a prospect on any need until he or she feels that need. The object of all marketing is *to create felt needs*.

What we really are discussing here, therefore, is the potential of persuading a prospect to regard what you offer as a better way to satisfy some basic need. We may speak of a customer feeling a need for a television when it became available, but the need was not for the TV itself, because it hadn't existed before. But the need for what the TV does or provides, a better form of home entertainment, did exist.

When we refer to felt or created needs for some product or service, we really are referring to a more fundamental need that the product or service is purported to satisfy in a better way. This is important to understand, because, as you will see, it is the basis for all successful marketing and selling activities. It is the key to motivating prospects to become customers, because it is the key to understanding the most basic wants and needs.

Granted, there must be a latent want in the recesses of the customer's subconscious mind before you can arouse it and create a need, which is a conscious want. If that latent want does not exist, you will not be able to create a need. Marketing and sales appeals directed to the general public's wants, conscious and unconscious, are those wants that most of us or, at least many of us, have. Most of us want to be happy, to be loved, to be prosperous, to be secure, to be warm, to feel comfortable, to be admired, to feel successful, to be trouble-free and to enjoy many other warm feelings and good things. It's safe to appeal to any of these. While we may feel them, most of us do not believe we will satisfy all of them, much as we would wish to. We tend to put many of these ideals in the realm of the desirable but probably unattainable. That itself leads to a skepticism that we often encounter in selling, and it is one of the obstacles to be overcome. The real "secret" of marketing success lies in how to make the prospective customer believe that any or all of these ideals is attainable. If you convince the prospect that he or she can attain one of these goals, you have created a felt need and guaranteed a sale.

The Marketing Process

Now let's look at marketing from the most pragmatic viewpoint: the specific steps in the marketing process. It is important to think of marketing as a process that includes a number of steps or progressive stages to advance from the offering to the sale.

Marketing is inherently an inefficient activity. It depends on a tiny fraction of success for whatever sales it accomplishes. A major corporation might broadcast its continual—almost continuous—sales appeals to the entire U.S. population, some 250 million today, to win perhaps one or two million customers, a return of less than one percent. But even that level of success is misleading. If the advertiser runs ten advertisements to win that less-than-one-percent return, the return dwindles to less than one-tenth of one percent; and if the advertiser runs 100 or 1,000 advertising appeals, well, you can see where this is going. The statistics of sales effort versus sales made can be depressing. The challenge to marketing is to operate viably despite the statistics. If you make an appeal to 100,000 people and the cost comes down to as little as one cent per person, the total cost is still $1,000. That is a lot of money for a self-employed individual. If you win only ten customers and sales as a result of that marketing effort, each resulting job has cost you $100, and burdens the job to that extent immediately. If the job is worth only $1,000, you have an immediate burden of $100 or 10 percent to add to your normal overhead. Those are the "facts on the ground." Every sale—acquiring each customer—costs you something. You always must be conscious of that.

As an independent contractor, you are not going to spend vast sums on frequent national advertising. You will have to make a sales effort, however, and the problem is the same: finding and selling successfully to those most likely to want what you offer.

For any given product or service offered, only a relatively small number of true prospects exists. If you are selling a rug-cleaning service, for example, you must address your appeal to those who are within your service area, who have rugs that need cleaning and who prefer hiring someone to do the job, rather than undertaking it themselves, or who can be persuaded to prefer hiring someone to do the job.

The first part of the challenge is fairly easy: Most of us who live in modern homes have rugs. A large percentage of those who have rugs probably have at least some that are in need of cleaning. If you can find them, you can try to persuade them that it is better to hire you to do the job than to do it themselves.

You can make several arguments:

- You might argue effectiveness of the cleaning process: You are the professional rug cleaner, so you almost surely can do the job more quickly, more efficiently and more effectively than the prospect can do it.

- You can argue costs: It might be difficult, but possible, to persuade the prospect that you can do the job more inexpensively than he or she can, but perhaps you can do it.
- You can argue convenience: You could make a good argument that having you do the labor saves the customer time.
- You can argue dependability: You can make a good argument there, based on the fact that you are a professional rug cleaner, you use the most modern and sophisticated equipment and you guarantee absolutely to get all the dirt out.

Prospecting

Of course, even having found that segment of the population that has rugs that need cleaning, you are faced with the task of persuading them to use your service. You already have carried out the early steps of the process:

- Deciding what you will sell
- Deciding to whom you will sell—finding the right prospects for what you want to sell

Suppose now that 0.01 percent of the general public is interested in your products, and 1 percent of that tiny group is likely to buy it. You are getting a 1 percent rate of success in selling to your true prospects. Wouldn't it be more efficient to address your appeal only to that 1 percent, the true prospects, instead of wasting your efforts and money on the remaining 99 percent? Of course it would. Marketing helps you find those prospects and make your sales appeal to them, rather than to the world at large.

The goal, then, is to find some means for singling out true prospects for whatever you sell and avoid wasting your efforts and money on the others.

You may want to ask, then, why some advertisers spend many millions of dollars on sales appeals to the general public. For many offerings, almost everyone is a prospect. Almost everyone uses certain kinds of products and services, such as long-distance telephone calls, and virtually every household uses soaps, detergents, colognes, soups and myriad other commodities. It doesn't pay, when selling common commodities, to make the effort to ferret out special prospects for commodities when one-half or more of the population are prospects. It costs money to do prospecting; for some kinds of marketing, it is not worthwhile to spend money seeking out prospects before launching a sales or advertising campaign.

That is not likely to be the case for what you sell as an independent contractor, however, so we will not dwell on this. In fact, yours may well be a rather special case: There is an excellent possibility that you do not sell to the general public at all, but only to a specialized segment of the public, those who contract or subcontract to independents such as you. Then again, you may not be making your offers to the general public at all, but only to organizations and perhaps only to prime contractors, from whom you hope to win subcontracts. In this latter case, prospecting is very much a must for you. It would be highly impractical to strike out blindly, hoping to strike the right organization or prime contractor by chance.

In short, then, your marketing program should include prospecting as a distinct preliminary and prerequisite to your sales campaign. You must find the right prospect for whatever it is that you wish to sell as an independent contractor.

4

Prospecting: The Search for Future Clients

In the pioneering days of the American Old West, prospectors were people in search of gold, about which some romanticism has grown up. Today's prospectors in the business world, while not romantic figures, are also people in search of gold, although of a somewhat different kind.

Prospects Cost Money

Regardless of the field and kind of service or product you sell, the marketing of your independent contracting service is rarely a one-call proposition. That is, you rarely win clients by the initial contact and presentation. It normally requires at least two steps, and more often more than two steps to close each sale and win a contract. In the most general of terms, the two-step process involves identifying a prospect—a prospective client—and then persuading him or her to become a client. This need for two or more distinct phases or steps to win contracts makes marketing of contracting services an expensive proposition. That inevitably influences the types of marketing methods you use to bring some efficiency to your marketing and over which you must exercise some degree of control. It is sensible, then, to establish as a basic marketing premise in this milieu of independent contracting that you win clients first by a specific program to find suitable prospects, and then a program to follow up specific sales presentations to convert as many of those prospects to clients as possible.

It is possible that you will encounter someone with whom you can chat, present your qualifications and close, all within the initial contact.

But such successes are exceptions or accidents for most kinds of contracting, and you cannot base your prospects for success on exceptions.

That does not mean that you ought not to try to close every prospect in your initial contact. Certainly, you always should try to close on the first contact with a prospect. Occasionally, this will work; more often, however, it will not. When it doesn't, you should probe further to find what it will take to win this client, perhaps even to judge whether he or she can be won and is a good enough prospect to merit a further investment of time and money.

Prospecting Expenditure Is an Investment

Every step in marketing costs you something, so there is cost involved in prospecting. Marketing typically is the most costly function of most business ventures, including independent contracting, and even that first step of finding prospects can be expensive.

Money spent in the search for prospects will appear on your books as marketing cost, yet the money spent is also an investment, producing an asset.

Prospects are always assets with specific market value that is in some proportion to the estimated sales value of the prospect. A good prospect for a $25,000 contract is at least nominally of greater value than a good prospect for a $5,000 contract. And so prospects can be graded in terms of their value. But there is another scale on which prospects may be valued: It is by the probability that the prospect can be persuaded to become a client. Some factory workers might be good prospects for a computer seminar, but most would not be; therefore, if you are offering such a seminar, a list of factory workers would not be of great value to you. Of even less value is a list of prospects about whom you know nothing that will help you decide whether they are good prospects for whatever you sell.

Buying and Renting "Canned" Prospect Lists

You can build your own prospect lists; in some cases that may be the only way you can get suitable prospects. In others, you may be able to buy or rent prospect lists that are right for you. Commercial mailing lists, for example, are one kind of ready-made or "canned" prospect lists abundantly available in a wide variety of kinds, classes and configurations, some of which are possibly of use to you.

Most mailing lists offered on the open market are the property of some business organization, although you usually rent them from a broker who manages and markets the lists for their owners. Typically, they are lists of the owner's customers or subscribers. Here are just a few of the many typical lists:

- North Shore Animal League lists of donors and mail order buyers
- *Guideposts* subscribers
- Spiegel catalog customers
- Bloomingdale's mail-order buyers

The owners of such lists rent—not sell—those lists at some fee per 1,000 names, usually, often with a minimum of 3,000 to 5,000 names. You must pay the fee for a one-time use, although whomever buys from you becomes your customer and you then own that name and address. Some list owners rent their own lists directly, but most put them into the hands of a list broker, who compiles the lists in vast databases and markets the lists for their owners, earning commissions in doing so.

Some brokers also offer lists, compiled from various directories, such as telephone books and membership directories. Here are a few such lists available:

- Automobile owners
- Spanish-speaking mail-order buyers
- Computer owners
- Canadian mail-order buyers
- Cookbook buyers

Many compiled lists are sold on an unlimited-use basis. That is, once you pay the fee, you may use and reuse that list as often as you wish without paying additional fees. You do not, however, own the list, so you cannot sell or rent it to others. You might, for example, be able to get a compiled list of all tool shops within 25 miles of your office. You can use that list any way you wish for your own marketing, but you cannot make it available to anyone else.

Some list brokers will even compile a custom list for you. You will then own that list and be free to do with it what you will, since you paid to create it.

The values of these kinds of lists vary according to what is known about those listed and how up-to-date the information is. If a list is simply that of individuals of a given profession or given locale, the list is not as valuable as is a list of prospects known to be buyers of some certain kinds of items or of some method of buying, such as mail order. The value of the list is proportionate to the kind and amount of demographic data on the list. This data might include income levels; how much prospects spend regularly on given commodities or via some buying medium, such as mail, electronic bulletin boards or shoppers' clubs; economic status; where and how they live; educational levels; the size and nature of their families; and sundry other details. All this information can be used to assist sellers in targeting prospects.

As with all prospecting, it is important to target your efforts. Demographic data that is of great value to a seller of securities may not be worth much to you as a contractor if you cannot use that information to enhance and improve your own marketing efforts.

The more information one has about the people on a list, the more selective and valuable the list can be. For example, if you sell golf equipment and supplies and you can get a list that is of affluent people who play golf every week and in most major tournaments and who are frequent buyers of golf clubs and accessories, that would be a highly desirable list and a quite valuable one. Conversely, if you could list the characteristics you would think ideal for a prospect list and someone could select such a list, it would be a valuable list. Suppose, for example, that a list broker has an extensive database of executives of all kinds and has, among other items, information on which of those on the list are ardent golfers. The list broker can then presumably pull a list of executives who play golf, and that list would be a good one for you because you sell golf equipment and supplies. If the broker can refine that list further and pull from it a list of those who are known to be frequent buyers of golf equipment and supplies, the list would become even more valuable.

Most brokers publish catalogs listing the many kinds of lists they can assemble for you. In effect, that gives you a large degree of power to specify just what you want on a list, within the limits of the list broker's ability to satisfy your specification. That enables you to target your market much more closely than you might otherwise. Targeting markets is an idea not always attainable to the degree that you might wish, but it is always desirable.

The primary customers for those lists are mail-order or direct-mail sellers. To them, the mailing list *is* the list of prospects, and they try to get orders from those listed. To others whose businesses are, like your own, not one-call businesses, those lists are useful to search out prospects. An automobile dealer, for example, might use such lists to appeal to people to visit the showroom to see the new models or to register their names for a lottery.

You may find this a practical idea for your own purposes. You can get mailing lists grouped according to many known characteristics—all bank presidents, all engineers, all suburban homeowners or all bridge players, for example. When you are in a field in which it is not practical to sell by mail—automobile or real estate sales, for example—the mailing list is not truly a prospect list. If you use the typical mailing lists of individuals, you would use them as lists of "suspects," individuals who *might* be good prospects for you. If your service is such that they are of greatest use to businesses and industries, a suspect list will not be of great aid to you. If, however, machine shops are a good prospect for what you offer, you can probably find a list of them within whatever you have decided is your service area.

In my own case, working as a consultant in marketing to the federal government and offering several services in that connection, I wanted lists of companies who contract with federal government agencies as a substantial part of their business. I did not have great success in finding suitable lists offered by the established list brokers. The lists offered simply were not what I wanted or needed. I felt that I was forced to resort to other means, and did so. In the end, I developed my own mailing lists in marketing my direct consulting services, proposal-writing seminars and especially my newsletter.

Qualified and Unqualified Prospects

The word *prospect* is tricky in its meaning; we use it to refer to those whom we think may be prospective clients only because we know next to nothing about them. When we learn a bit more about them, we are better equipped to judge which are possible prospective clients and which are not likely to become clients.

When you have a list about which you know very little, anyone could conceivably become a client, so you must consider him or her a prospect, for the time being. If you know a little about the people on the list, you may be able to make a reasoned judgment. I know without asking that the average student is not a prospect for my services as a proposal consultant. I would therefore not address a list of students as possible prospects for what I sell. On the other hand, if I had a list of marketing managers, I would know that some of those probably would be prospects for my services. In a sense, then, they are all prospects, at least until I can prequalify the list and screen out those who truly are not prospects for me. For example, marketing managers for candy manufacturers would not be good prospects for me. Candy manufacturers may sell to government and other large organizations, but they rarely use proposals to do so. It is not in the nature of their marketing. Marketing directors of engineering companies, on the other hand, are *possible* prospects for proposal services.

If I have a list of 10,000 names and 2,000 of them are names of marketing directors in engineering companies, do I have 2,000 good prospects? No, I have 2,000 possible prospects, and I must regard them as so, because I cannot call up or call on 2,000 marketing directors to try out my sales presentation and see which will be responsive. To be practical and to avoid wasting my time trying to sell those who are not true prospects for me, I must do some kind of screening to eliminate those who are not prospects, or qualifying to identify those who are prospects. *Qualifying* is the more popular term used to refer to measures taken to evaluate a prospect and judge the wisdom of investing in the effort to make the sale.

In any case, we must recognize that even given a list or access to a large number of people who ought logically to be prospects, it usually is necessary to go at least one step further to qualify the prospect.

Prospecting Methods

Useful as it can be, the commercial mailing list is only one of many ways to find prospects. This ready-made list of prospects, is probably the only ready-made list of prospects you are likely to find. Unless you are fortunate enough to get lists that are just right for your needs, prospecting for good candidates for your services means finding them on your own by whatever means you can discover or devise. Let's look at some other prospecting methods: seminars, trade shows, networking and image enhancement.

Prospecting by Seminar

Many independent contractors, myself included, have found the seminar a highly productive way of prospecting. Almost without exception, every seminar I ever presented to a general audience of registrants produced one or more leads that led to contracts.

I did not employ seminars as marketing efforts initially. I saw seminars as part of the services I provided interested users. My independent contracting and consulting service covered writing, counseling, problem solving and managing the entire proposal process for my clients, even designing or helping design the systems and programs offered by my clients to their prospective buyers. My seminars offered as much of all those services as I could deliver from the dais: I taught contracting and proposal writing, and responded to questions to the best of my ability.

One might think that having taught as much of what I know as I could deliver in six hours of lecturing, I would be reducing the number of clients who needed my help. In fact, the more I revealed, the more listeners perceived there was to know about the subject, the more they wanted, and the more likely they were to see the value in retaining me to help them with a specific proposal.

Many entrepreneurs offer free seminars, although too often the free "seminar" then turns out to be a marketing presentation or, less elegantly expressed, a sales pitch. I think that is largely self-defeating because it cheats those who attend, expecting to get at least some useful information along with the marketing presentation. If you choose to offer a free seminar, it probably will work best if you deliver some useful information before you enter into the sales effort. The purpose of the seminar is to attract those who are likely to be prospects for whatever you offer, to demonstrate your own ability to provide a useful service, and to develop a final list of prospects worth following up.

There is still another way to use seminars in your prospecting. Many seminars are run by others who use a series of several speakers, each making relatively brief presentations. There are often literature tables at such seminars, and if you are a speaker, you can leave your literature on the table, and introduce yourself from the dais.

Many companies organize an in-house speakers' bureau of its executives as a public relations asset. They then provide free speakers to clubs, schools, associations, seminars and events. You can do this yourself, on the small scale of the independent contracting venture. Make up a brochure that describes topics you will speak about and send it out to relevant associations, business clubs, local newspapers and wherever else you believe you can find a suitable audience.

Conventions and Trade Shows

Annual conventions of relevant associations offer good prospecting opportunities. Many offer seminars, for example, and you may be able to work something out along those lines. But many also have exhibit halls or trade shows, and staffing an exhibit booth at such an event is a way of gathering a prospect list. There also are opportunities often to distribute your sales literature and to move about and meet a great many people in a kind of networking mode. You also may be able to get a membership list and put that to good use.

Networking

Networking is a particularly appropriate prospecting method for many independent practitioners. In principle, it consists of making yourself and what you offer known to as many people as possible through personal contact and resulting word of mouth. Some have well-organized and highly detailed formal programs for networking, but that is not an absolute necessity. Some of what you have just read is networking. Moving about at conventions and making yourself known to as many people as possible is one method. Joining associations and becoming active in association affairs is another.

Raising Your Image

All of these suggested approaches are intended to raise your image, making you as visible as possible to as many potential clients as possible. That is the central idea. You can further enhance it by maximizing your writing and speaking activities in circles where they will reach the eyes and ears of those who are likely to be interested in what you offer.

Other Considerations

A number of other considerations can have a direct impact on prospecting. Some of them are arbitrary choices you make, while others are circumstances over which you have no control. All are important to consider in making decisions as to your prospecting activities.

Defining Your Service Area

By now, presumably, you have decided just what services you will offer and what kinds of prospects you think are most likely to become your clients. Because independent contracting covers a wide field of possible industries and services, we must discuss the subject of defining prospects and clients in rather broad terms. Defining your service area is another matter; we can get quite specific in that.

You have several choices in the matter of a service area. You can decide to accept assignments locally, within your state, nationally or even internationally. International contracting is not unusual these days: The world has become a much smaller village than it once was, and people jet around the world quite casually on business and pleasure. A large number of independent contractors do accept overseas contracts, despite being small and even one-person ventures. Many hurried to help clean up the problems in Kuwait after the end of the Gulf War, for example, as they had earlier in the case of Vietnam and other places in the world.

Trade-Off Considerations

Some factors may dictate your decisions. You may need to define your service area, for example, based on whether there are enough prospects for your services within your city and suburbs to support your practice. If not, it may be necessary to have a wider service area. There are other questions: How wide a service area is practical for you to meet client needs efficiently? Will clients in faraway places wish to engage your services? They would, presumably, have to pay your travel and related expenses; would the client consider this practical or acceptable? Are you willing to take long trips to deliver your services? Would you require clients to pay for your time in travel? Would this cost affect your ability to be competitive? Does what you do depend on cost competitiveness?

The concerns and decisions regarding your service area are thus not independent of other concerns, but may be entwined with them. You may start with some defined program but later find it necessary to expand, add to or otherwise modify your service offerings. You must try to be flexible, especially in conducting what essentially is a custom service, and react appropriately to conditions you encounter. In my own case, I started out to be a writer, reacted to conditions and opportunities to become a consultant along the way and eventually became a public speaker and instructor, as well. Finally, I used the experience and knowledge I had gained over the years to become the author of many books on business subjects.

Prospects for Contracting or Subcontracting?

The nature of what you do or how you choose to do it is another factor that could define the most suitable prospects for your services. Some independents choose to be contractors, others choose to be subcontractors and still others seek and accept both arrangements. The prospects for prime contracts may be and often are quite different than the prospects for subcontracts, and thus may dictate the nature of the prospecting campaign.

The typical practices in the business or industry you serve may dictate some of your choices. If you are contracting in such industries as construction, focusing on major projects, you usually will be seeking subcontracts. That usually means registering your name and type of work you do with all the construction companies in your service area, as one early step. You also will want to be alert for all projected construction jobs in your area, as construction companies from distant places may bid and win the prime contracts and arrive to do the work, often preferring to hire their subcontractors locally when they can.

In many other fields you can pursue prime and subcontracts as opportunities arise. As a contract writer, I won many prime contracts with clients to do various kinds of writing jobs for them, but I also won many subcontracts to provide writing services for larger firms running the prime contracts.

Again, market research will help you determine who needs your product or service. Who needs and buys such services? How do they usually buy them—as individual items for which they contract, as part of large contracts or both? Do they buy from independents or only from major suppliers? Of course, this assumes you are studying all the current practices and conventional sources of business in trying to identify your best prospects. Must you confine yourself to those established avenues of trade? Is it possible to uncover or invent other avenues and other prospects for what you do than the traditional ones? Must you follow the well-worn trails of the past in tracking down the most promising prospects or is it possible that you may blaze new trails and find new classes of prospects?

Imitation or Innovation?

The late John D. Rockefeller, the millionaire industrialist, advocated as a success strategy the observation of what someone was doing with great success and imitating him or her. It's a common enough approach. One man I knew and was employed by started a subcontracting technical-writing business at his mother's kitchen table, and expressed the ambition to be as big as a small firm in that business that was then doing several million dollars worth of business each year. When I last had occasion to hear news of his firm several years ago, he was doing more than $500 million each year and employed several thousand people.

There is nothing wrong with imitating someone successful, trying to do what that other party is doing better than he or she is doing it and achieving even greater success than the other party. A great many successful businesses are built this way. But there is another way. Many highly successful businesses are built by doing something no one else is doing: by innovation, rather than imitation. The basic idea that inspires most highly imaginative people, the principal innovators, is that there is always a better way to do everything; every function, system and product can be improved. Make that your mindset and you will condition your subconscious to be questing always for the better way.

The Meaning of Innovation

As a business concept, innovation is multidimensional. In general terms, it means being different and doing whatever you do differently than anyone else does it, it may mean bringing an entirely new and different service or product to the market—or it may mean all of these things. The first writer who ran advertisements offering to write personal letters for individuals who needed that help was an innovator. She perceived what she thought to be a need, and she put the idea to the test, with great success. The first contractor who offered to supply office temporaries was an innovator, too, as were the brothers Bloch when they established the first of a national chain of offices offering income tax services.

Innovation does not necessarily require revolutionary or highly dramatic change. As an engineer might put it, the goal may be evolution, not revolution. Even a small change can produce dramatic results. Any change that produces a demonstrably better result is an innovation that may itself be the prime basis for success.

Innovation means new or newly introduced, but *new* is itself a relative term. That means that an innovation may be a simple rearrangement of familiar things. Cartoons were not new, nor were the problems of writing effective maintenance manuals for highly complex military equipment, but combining the two—writing maintenance manuals in cartoon-book form—was an innovation, and the military his since written many manuals in that style.

In prospecting for clients, innovation can be important. It can affect your prospecting in several possible ways, especially in devising and using new ways of seeking prospects or in seeking a specialized or niche market.

How Innovation May Come About

Innovation can be the result of earlier experience, during which you devised more effective ways of doing whatever it is that you do. It may be an accidental discovery of some useful idea or product; or it may be the result of deliberate and determined effort to find a better way. That may in turn be the result of simple dissatisfaction with established practices and lengthy experiments with alternatives to discover the best ways, such as is practiced in the formal value analysis methodology. I have more than once found that the reverse of an accepted practice produces better results.

Innovation Must Be Beneficial

Innovation is not necessarily a boon. The innovation that you consider to be an improvement is of interest at this point only if it offers you some distinct advantage in prospecting. It must help you, directly or indirectly, in finding and closing good prospects. That is, it must be of itself an advantage in the prospect's eyes or enable you to offer prospects some advantage, such as lower cost, better results or faster results. That helps you find prospects, as well as close deals, because it offers some benefit with which to attract prospects, to draw their interest and inquiries and enable you to make a presentation. On the other hand, it may be a better way to search for, find and qualify prospects.

5

Marketing and Sales Tools

*Better mousetrap or not, the world will not beat a path to your door to press
their dollars into your hands. You must go out and woo the world if you
want to win enough clients and contacts to build a business and to prosper.*

We Are All Hucksters

Many in the advertising industry, commonly referred to as "Madison
Avenue," refer to themselves as hucksters. This term was originally
associated with street peddlers or "hawkers," from which word it derives. I first became aware of this word as a term for those men who
walked down the narrow alleys behind the rows of houses in my native
Philadelphia, carrying vegetables and fish, shouting loudly: "Fresh
fish!" "Cantaloupes!" "Tomatoes!" or whatever wares were in the baskets
on their shoulders.

Of course, you cannot march down the street shouting, "Contracting!" "Home improvements!" "Newsletters written!" "Exhibits built!"
Still, that is what we must do, in effect. We somehow must sell whatever
it is we offer to any and all who can use our services, and that means
taking the initiative in seeking buyers. You will starve rather rapidly if
you wait for the world to seek you out and beg to buy from you, despite
some popular platitudes to the contrary. You must work to persuade the
world that it needs what you have to offer; you need to sell it. Constantly.
Everyday.

We tend to use the word *marketing,* as a euphemism and perhaps a
more dignified word than *selling* or *sales.* Certainly, it is a more dignified

word than *huckstering,* but huckstering really is what we do when we market. There are theoretical and technical differences between marketing and selling, and certainly between marketing and huckstering; still, there is no denying that the end-purpose of marketing is to sell. Let us therefore use the terms *marketing, selling* and *sales* more or less interchangeably, as most of the business world does.

We must, of course, turn to the most modern methods of huckstering. The big corporations do their huckstering today with huge and costly magazine and newspaper advertisements, with million-dollar contests and, especially, with elaborate and costly TV commercials supporting a variety of TV programs. They spend many millions—billions, collectively—to shout their modern-day versions of "Fresh fish!" and "Juicy cantaloupes!"where millions of people can read, see and hear them. Shouting our wares on that grand scale is out of reach for us small folks, independent contractors and small businesspeople that we are, but we can turn to our own versions of the ancient art of huckstering on a scale suitable for us. And to keep our dignity intact as serious entrepreneurs, we shall henceforth refer to our means for shouting, "fresh fish!" as sales or marketing tools.

Modern Huckstering Requires Modern Tools

Every business requires marketing and sales tools. In general, that may include brochures, sales letters, proposals, catalogs, item descriptions, specifications, signs, decals, posters and a wide variety of presentation materials, such as movies, slides, slide tape and filmstrip programs, and sundry other items. In addition to those, there are the tools of public relations, also used for marketing and special sales promotions, including press releases, press kits and even speakers' bureaus, with associated presentation materials. In short, marketing and sales tools include a wide variety of items, including the software and hardware needed to carry out whatever marketing, sales or special promotions you need to accomplish your marketing.

Of course, only the largest organizations would be likely to have need or use for all of the tools listed here. The ones that apply to your own situation are going to vary, depending on a variety of factors, and are likely to be relatively simple. Those sales tools that you are most likely to need as a minimum, and which are most likely to be practical ones for you, at least in the beginning, are sales letters, brochures and proposals, although you may also be able to make good use of press releases. The latter are probably the easiest and least expensive to create. Beware, however, that word *easiest*. It is easy enough to write a release, as it is easy to write a sales letter, brochure or proposal, but simply writing these items is not the same as writing *effective* ones. The vast majority of all of these sales tools that come into being are not well done from the viewpoint of sales effectiveness, even if they are artistically well done and aesthetically pleasing. Prospects may admire your brochures and other tools you develop, but the only admiration you really quest for and can benefit from is the conviction that impels a prospect to become a client: to buy what you offer. This is the only kind of approbation you can put to good use and the only valid measure of quality in reference to sales tools. Advertising and sales history is full of examples of presentations that the public loved but were not persuaded to buy. The distinction is critical.

Beauty versus Utility

Despite common misunderstandings, aesthetic quality does not equal utility. Business owners developing sales materials are far too often guided in their choices and decisions by what they believe to be pleasing layouts and eloquent language. Nor are business owners the only ones guilty of this mistake. Too often, professional experts including copywriters and advertising specialists, allow themselves to be lured by aesthetics and thus neglect the practical considerations of motivating prospects to buy—to become clients.

What tools you need to sell your services as an independent contractor depends on the kinds of services you wish to provide under contract. To start, we must divide the markets for independent contracting into two broad segments: those that require more or less formal contracting and negotiation to close sales, and those in which sales can be closed spontaneously, often without preliminaries or other formalities, and often requiring only a handshake—i.e., an informal agreement. The entire methodology of marketing and closing is different for each of these two broad classes.

The division primarily represents the size of the typical contract. If the kind of work you do as a contractor normally entails small jobs and short schedules—e.g., costing not more than three figures and involving only a few hours' work—you will probably close most of your sales with entirely informal agreements, usually with nothing more than correspondence to attest to the trade. On the other hand, if your typical sale involves effort of many days' or weeks' duration and dollars well into four or more figures, you surely are going to execute written agreements and quite possibly formal contracts. If you do not insist on it, your client probably will.

The size of your typical project dictates what kinds of marketing and sales efforts are necessary to close the average sale and thus what sales tools you need. Even the major corporation signing what is a small contract to them, such as $25,000, is likely to insist on a written agreement of some kind to document the obligations of both parties. That is not necessarily a multipage contract, however; it may be a purchase order with a clear specification attached to stipulate the agreement or a simple letter of agreement, a common form of contracting for small jobs.

Many kinds of independent contracting fall between the extremes. I presented many seminars and other single day consulting services for $1,000 or thereabout, with expenses, and generally used an informal letter agreement. Certainly, it was not worth our time to spend many hours in negotiations for contracts of such brief duration. In most cases, we exchanged correspondence and the work was authorized by purchase order or signature on one copy of the correspondence.

There is one other consideration here. That is the nature of your typical clients—i.e., your target market. Selling to private citizens is a much different proposition in many ways than selling to businesspeople, because the motivations are different. You must understand those motivations if you are to market effectively. If you are a home-improvement contractor, for example, you may be selling to individual homeowners or to businesspeople—builders or perhaps dealers of some sort. The home-owner is concerned with and motivated by considerations of personal satisfaction, comfort, pride and convenience. The businessperson is concerned with and motivated by considerations of costs, profits and keeping his or her own clients satisfied. That makes quite a substantial difference in your needs for sales tools and the nature of your entire marketing program. Both types of client may be the buyers of small- or large-budget services. Neither type of client is one you are likely to close on a first call, perhaps, but in general it is likely to require a more extended and more intensive effort to close a sale with the businessperson than with the private individual.

Therefore, while we will discuss a variety of sales tools here, all the tools that all average independent contractors are likely to need, recognize that some may not apply to you and consider only those that do apply to you and the kind of contracting you do or plan to do—to, that is, the sales and marketing needs you have or are likely to have.

Sales Correspondence

Sales correspondence is sent out to support any phase of the sales effort, including initial prospecting for leads, following up, responding to prospects' and clients' requests and closing. *Sales correspondence* therefore refers to a fairly broad and general class of sales materials, including but not restricted to sales letters, bids and quotations, letter proposals and related items.

Sales Letters

Not all sales letters are direct solicitations; some are responses to inquiries and are intended to encourage sales through low-key selling. However, for the purposes of this discussion, we are concerned primarily with letters that are sent out, perhaps in quantity, to prospects who already may or may not be established clients. They may be preprinted form letters, and they may be individually addressed and typed, so that they do not appear to be printed form letters. Actually, with modern computers and mail-merge software, form letters easily can be sent out in quantity while addressed individually to prospects as though they were individually typed.

If you have taken the time to open and read your morning "junk mail" occasionally, you no doubt have encountered the typical frantic sales letter used by many direct-mail copywriters. It shouts excitedly at you that the moment is now to seize the opportunity to buy the new plastic frying pan at a sensational low price with an unbelievably ironclad guarantee, and a half-dozen assorted utensils thrown in as bonuses for prompt action. A subdued example of such a letter is shown in Figure 5.1.

This kind of sales letter is the written version of the late-night TV commercial or infomercial and the boardwalk hawking of miracle potato peelers and glass cutters. Such letters do work to sell potato peelers, glass cutters, music tapes and discs and many other commodities that people will buy on impulse—without lengthy deliberation and family councils to reach agreement on major expenditures. They don't work well, however, in selling contracting or other services of a more serious nature, which mandates the need for serious study and negotiation and serious consideration of costs.

That is not to say that you ought never to use a sales letter. A sales letter is the centerpiece in many types of sales presentations, introducing the seller and the service or product. By itself, it may produce small sales, but it cannot produce large sales and large contracts, and should not be expected to do so. Its primary usefulness is as a prospecting tool for initial contacts and follow-up, in some cases. Figures 5.2 through 5.5 are examples of sales letters more suited to your probable needs.

Sales Letters in Pursuit of Leads

When you write the sales letter in pursuit of leads, you must remember your objective: You are not trying to make sales; you are trying to develop sales leads, which means finding those individuals who display some active interest in what you wish to offer, enough to merit follow-up sales effort on your part. Therefore, you must persuade the respondent to do something to exhibit that interest. That is, you are trying to "sell" the reader a serious interest in learning more about your offer.

Sales letters in pursuit of leads normally ask the respondent to take some action, which would establish the respondent as a serious prospect. The respondent in such cases is asked to call for an appointment, come in for a visit, request a copy of a newsletter or brochure or otherwise display active interest. On the other hand, many sales letters are in pursuit of direct sales and ask the respondent to order then and there, as in Figure 5.1. It is not likely that you will have use for this kind of letter, so we will merely recognize its existence and go on.

Sales Principles

The first and foremost principle of all selling efforts is that of showing the client why it is in his or her interest to listen to and consider your offer. A sales presentation should be an offer to do something rewarding or beneficial for the prospect. People want to make more money, save money in what they buy, get ahead on the job, be attractive, be liked, be respected, be loved, be secure and be part of something, to name just a few normal human desires. And, as already noted earlier, people in their business lives have the same desires as people do in their private lives, although their desires are focused a bit differently. The letter of Figure 5.1 offers a golfing enthusiast a bargain price for a new set of golf clubs. The next letter, in Figure 5.2, makes an emotional appeal to the desire for summer comfort, while also offering savings. Note that both letters offer evidence to back up the promises of cost savings and do not ask the respondent to simply accept an assurance of savings. Note these factors also in the remaining sales letters, Figures 5.3 through 5.5, exemplifying

FIGURE 5.1 Typical Direct-Mail Sales Letter

Dear Colonel Hill:

Due to a fortunate purchase we can offer you a sensational bargain: a 24-piece set of genuine BackSwing golf clubs, as illustrated in the enclosed brochure, for only—are you ready for this?—$189.90! (Yes! That's right! Only $189.90 for all 24 pieces.)

Why so low a price for this sensational item? Simply because of a factory overstock. The Outdoor Sporting Goods Co., who manufactures this line, was unexpectedly unable to sell as many as it usually does in Europe due to a sudden change in the exchange rate, which priced it out of the European market and left it unexpectedly with more of these fine sets than it could sell. We persuaded the manufacturer to let us have these outstanding golf clubs at a phenomenal discount by agreeing to buy the entire stock. And now we are passing the savings along to our valued customers.

We don't know how long this stock will last—it was limited, to begin with—and there are no more when the current stock is gone. Undoubtedly, they will go rapidly at this price, so we urge you to act promptly. A telephone call will hold a set for you for a few days, but you can be sure of getting a set by mailing in the postage-paid order form enclosed or calling your order in by telephone or fax.

Don't miss this opportunity. Another like it may never come along.

Cordially,

H.T. Bartlett
Sales Manager

appeals for several other kinds of contract services. Note the variety of services one may offer as an independent contractor.

Note also that the desires to which promises appeal all have some emotional content. You may be sure that without exception, or with only

FIGURE 5.2 Sales Letter Seeking Leads for Home-Improvement Contracts

Dear Ms. Williams:

It's time once again, with summer almost here, to think about summer comfort. I am delighted to be able to offer you a modern sunroom addition to your house at nearly 35 percent less than the normal cost of such an addition—if you act within the next 30 days.

The reason for the lower cost is simple: Labor is by far the major cost in construction today, and we have reduced our labor costs by nearly one-half. We do that by using modern prefabricated components. We thus find it necessary to do very little cutting and fitting on the job.

That saves time and litter too. We can do the entire job within four weeks, and you will hardly know we are there. First we come out and measure everything. Then we assemble all the standard components you need, and do most of whatever cutting and fitting is necessary here in our shop. Meanwhile we prepare the foundation to receive the components. Finally, we assemble and install the unit.

Just think: Even though this is late spring, you can have your new, breezy sun room in time for early summer. We arrange easy financing, and you'll be surprised at how little it costs.

A telephone call to the number above will bring me out to show you samples and answer any questions you have. But you must act now because our busy season is just beginning, and we can accept new orders only for about 30 days.

Call today. There is no obligation, of course. Ask for me personally.

Cordially,

John Martin
JM Construction Co.

FIGURE 5.3 Solicitation for Newsletter Contracting

Dear Ms. Berger:

More and more organizations are discovering the benefits of maintaining contact with their customers by publishing newsletters of their own. They shy away from the editorial duties, however, but they need not fear these: We are happy to handle these functions for you.

Enclosed is a sample copy of the bimonthly periodical that we prepare for the Grove Village General Hospital. It is typical of a number of periodicals we do for clients. The hospital supplies us with raw material for each issue, which we supplement with additional material we collect ourselves from various sources, as necessary. We prepare a mockup of the new edition and submit it to the hospital for approval. If necessary, we meet to discuss the new issue.

We supply the hospital with page proofs for final approval and, if necessary, make final changes.

Thus, the client is always in control of what is to appear in each issue, although we handle all the editorial work. In fact, we handle whatever elements of the entire requirement the client wishes us to handle. For some clients we handle everything, from writing to mailing; for others, we handle part of the process. Here, in fact, are the steps which you can handle yourself or have us handle for you:

- Original design
- Writing, editing, illustration
- Layout, page makeup, preparation of camera-ready copy
- Printing
- Mailing
- Maintenance and management of mailing lists and correspondence

Enclosed is a brochure. We shall be pleased to meet with you to discuss what we can do to help you create or manage a periodical of your own. Please feel free to call or write with any questions you may have or to request any additional information.

Cordially,

H.W. Higgins
Higgins Publications Services

FIGURE 5.4 Soliciting Leads for Contract-Labor Sales

Dear Ms. McLaughlin:

Downsizing is a fact of life today, a necessity for many organizations, large and small. It brings a significant reduction in operating overhead, but it does not have to mean a reduction in staff capability. It is our business to see to it that you do not suffer a reduction in staff capability when you downsize.

We are suppliers of temporary staff in a variety of categories, from file clerks to accountants, engineers, scientists and even physicians. All are highly trained specialists, ready to begin work to satisfy your needs immediately upon reporting in to you. There is no need for the lengthy get-aboard learning period you normally encounter with permanent new hires.

Important though it is, that is not the only advantage of using our services. You may ask for résumés, interview candidates and select those you want, but you avoid the recruiting costs, agency fees and severance costs. You can terminate any temporary staffer immediately and even abruptly when the need has passed or in the event you are not completely satisfied with the individual's performance.

I am enclosing some general information for you, but please feel free to call on me for any additional information you need. I would like to call on you personally to acquaint you with the special advantages we offer you in meeting your temporary staffing needs. You will find that our service is quite special.

Cordially,

Arthur Willoughby
Staff Support, Inc.

FIGURE 5.5 Solicitation Letter from a Meetings Specialist

Dear Mr. Prescott:

Organizing conventions, conferences, seminars and other meetings requires a great deal of tedious preparatory and administrative work, but it is not necessary to tax your resources to do all this yourself. I offer relief—a better way—by handling all of this for you. And perhaps more to the point, I specialize in the very special meeting, the unique and memorable occasion.

Among the meetings-management services I offer are these:

- Finding suitable sites and arranging for their availability
- Writing all necessary notices, letters and brochures
- Finding and arranging for speakers
- Arranging for all necessary furniture, fixtures and equipment
- Finding and arranging for or hiring all speakers and presenters
- Any miscellaneous other details

Of course, all duties and functions are carried out under your monitoring and approval so that you are in control, although not burdened with anything other than stipulating your wishes, monitoring my activity and approving or disapproving any action I take or recommend.

A more detailed description of the functions and services I provide is provided in the enclosed literature. I will be most pleased to discuss your needs with you by telephone or in person, preferably the latter, since I can then bring along a special presentation of meeting facilities that I have collected and organized especially for my clients. Most of these are not well known to most people so that you will see a wide choice of quite special locations and facilities for your meetings, many of which you did not even dream existed.

I look forward to an opportunity to meet with you personally to discuss your needs and suggest meetings that you and attendees will find most impressive and most memorable.

Cordially,

Walter Wilton
Meetings, Inc.

the rarest exceptions, we are all motivated to buy far more by our emotional drives than by our rational ones, although we are most reluctant to believe that. Reason does enter the picture to support the buying decision, but only as a support of the decision already made emotionally. The client needs to be able to rationalize the decision; hence the necessity for a rationale.

For example, the letter soliciting newsletter-contracting services, illustrated in Figure 5.3, bases its appeal primarily on the immediate advantages of publishing a newsletter, but also focuses on the advantages of having someone else do most of the work.

The letter in Figure 5.4, which solicits leads for contract-labor sales, bases its appeal primarily on cost reduction, but is careful to mention the convenience of hiring and firing temporaries, as compared with permanent staff. And finally, the letter in Figure 5.5, which solicits leads for contracting meetings management and services, stresses the offer of a special list of very special meeting places and facilities. Not only will the offeror take over all the headaches of managing the whole thing, but will make the executive a "hero" by leading him or her to a special and memorable meeting event.

That latter example is a special one: Who doesn't want to be outstanding, to be noticed as someone special or one who does something quite special? Take, for example, the public personality who earns $5 million a year, but complains that his or her opposite number in another studio or on another network gets $7 million a year. The difference in money does not really matter at that level of income, but the ego of the performer who gets "only" $5 million is bruised. The money is a way of keeping score, a mark of esteem, which is more important to such figures than money, although they can also easily rationalize that need as a practical one.

The truth of the emotional drive being paramount is illustrated every day in successful advertising. For minor purchases, where the risk of monetary loss is insignificant because the money involved is insignificant, the advertiser rarely bothers to introduce much of a rationale, but relies entirely or almost entirely on the emotional motivation evoked by the promise. Beer commercials are a good example. They don't really sell beer, of course, and they certainly waste little or no time discussing quality. Instead they focus entirely on selling good times, fun, which allegedly results from buying and drinking their product. It is that which is the benefit, clearly implied by the nature of the advertising.

Of course, as an independent contractor you are not likely to sell services or products to make your clients more beautiful or more loved, but your work, whatever it is, must produce benefits of some sort. They might make a client's home more beautiful or safer; they might save a client money or make his or her business more profitable; they might solve client problems and make his or her business life more pleasant and less hazardous; or they might produce any of many other possible benefits. It is up to you to reason out what those compelling benefits are and which are most likely to be the most motivational for any given class of prospect. Without question, those benefits and motivators are there to be found; your job is to point them out to the prospective client. It would be a gross mistake to wait for the prospects to make the effort to think out the benefits for themselves. You must find and express the promise in terms that appeal to and take advantage of the human tendency to satisfy their emotional desires as an important part of decision making.

Promise and Proof

It is appropriate to summarize the principles of business letter writing in terms of the two basic elements of sales, the promise and the proof:

- The promise is the benefit itself, or the offer. It is the promise to make the client wealthier, more secure, more beautiful, thinner and so on.
- The proof (or *evidence,* if you prefer that term) is the rationale that demonstrates the logic or validity of the promise and "proves" it in whatever terms the client will accept as proof. This latter provision is important, for nothing is so critical as the client's perception of fact or truth.

Offer the properly motivating promise, what your offer will do for the client and the proof that you can and will keep the promise (that your service or product will do what you say it will do for the client). Remember, again, that proof or evidence is whatever the client will accept as such. It need not be more than that, but it cannot be anything less either.

Be careful you don't dilute your letters's impact by promising the moon or an entire laundry list of benefits. The sales appeal ought to be built around one central feature, one major benefit that you believe will have wide appeal, that you believe you can back up or prove, effectively, and that will be credible.

Credibility

It is quite possible that your service can produce remarkable results. That does not mean that they are credible results. What is important in marketing is not always what is ultimate or objective truth; what is important is the client's truth, what the client believes to be true. Your proof will not stand up to a client's skepticism. If the client believes that all money-making plans are bogus, the best such plan will not be believed by that client, regardless of the evidence produced to support the promise of riches.

Thus you must make a promise that a reasonable person would find credible, and then concentrate on proving that promise valid. A modest promise that you can back up will be far more convincing than an impressive promise that you can't substantiate in some manner. Even if it were true, if the client cannot be persuaded to believe it, it might as well not be true. Thus you may find it necessary to gear your promise to the evidence you can produce.

Style in a Sales Letter

All letters, including sales letters, must be easy to read, have specific objectives and state specific facts in the clearest terms possible, if they are to do a proper job. Aside from that, sales letters must be based on reliable sales principles if they are to have even a faint chance of succeeding in their purpose.

Because these are sales letters and no effort is made to hide that fact, they do not have to follow the normal style of business letters too strictly. Many devices may be employed to make these sales letters command attention and stand out in some way.

One refinement you can make is to use a headline, as many do. For example, any of the sales letters shown here might have been enhanced with a headline (usually preceding the salutation). Here are some typical headlines that might be used in the foregoing models or other sales letters (and these headlines are also good examples of promises):

- Genuine BackSwing Golf Clubs at Bargain Prices
- Your Own Newsletter Without the Pain
- Solve Your Staffing Problems and Reduce Overhead
- Want Special, Memorable Meetings?

Persuasion is the principal objective of a sales letter, and the effort to be persuasive is often an uphill battle because it is often asking the party addressed to give up something valuable and not easily parted with, such as money, time or convictions. It is therefore critically important that a sales letter present information in a style that makes the information both believable and acceptable, essentials in any sales presentation.

Today's average adult is better educated than any in history in both the formal sense of schooling and the informal sense of learning from everyday experience. We are bombarded with information today from radio, movies, TV, newspapers, magazines and now from computers and public databases. "Hype," a colloquialism derived from the word *hyperbole* and meaning grossly inflated and exaggerated claims, with a strong suggestion of deliberate falsehood, shouldn't be a part of your business letters. It is still used, but it has limited influence, especially where you are seeking sales of size. The public has become far more knowledgeable and sophisticated today, and people tend to recognize hype for what it is, even when cleverly done. In fact, the fewer adjectives and adverbs you use, the more believable your letter will be. Stick with nouns and verbs, avoid superlatives and quantify as much as possible. In quantifying, use the most accurate figures possible and do not round them off. Making 937,682 into 1,000,000 or even "nearly one million" attacks credibility at once. The reader will believe 937,682 unquestioningly, but may smile indulgently or snort in disbelief at 1,000,000, assuming automatically that it is hype.

Use a direct style here, use imperatives and personalize the message as much as possible. Instead of, "Anyone who takes advantage of this offer can save a great deal of money," say, "Take advantage of this offer now. You will save hundreds of dollars." "You" (the client or prospective client) is always a key word, the most important word in the language. Use the client's name in the body of the letter, too, if you are using modern computer or word-processing technology, which permits you to do this even in a form letter, for example: "Mrs. Smith, take advantage of this offer now. And ask for me when you call. I will be delighted to help you personally."

Be especially careful in writing a sales letter to make it *appear* to be easy to read. Solid blocks of text and long, run-on sentences are formidable, so make your sentences short and direct, and break up your copy into frequent short paragraphs, with indents at the beginning of each and spaces between paragraphs.

Marketing wisdom accumulated over the years has demonstrated that certain words and phrases used as or in connection with promises of benefit are especially effective in their appeal to readers and apparently never wear out, probably because most readers want to believe these words and phrases. The following are just a few of those magic terms:

- Free
- Different
- Successful
- New
- Exclusive
- Efficient
- Discount
- Independent
- Guaranteed

Regardless of whether these captivating words are appropriate for your needs the implication is plain enough: they are emotion-ladened words, expressing ideas most of us respond to in some manner or degree.

One final caution: Don't allow the message in your letter to end abruptly and indecisively, hanging uncertainly in the air, at the end. Tell the reader what you want him or her to do, for example, come in for a personal demonstration, fill out the enclosed form, call and reserve one now, let us come to your home to demonstrate or other suggested action, as appropriate. If you fail to do this, many (probably most) prospective clients will do nothing at all. Most people need to be urged and directed to do whatever it is that you want them to do. Study commercial advertising to find examples of this principle at work.

Bids and Quotations

Those who request bids and quotations regularly in the course of their business (especially large firms and government agencies) and firms who submit bids and quotations regularly in their marketing, usually have their own standard forms for making up those bids and quotations. In fact, it is not at all necessary for any firm to design such forms of their own, for major printing houses and office supplies vendors can usually supply such forms, designed specifically for many kinds of industries and businesses. In the majority of cases, therefore, bids and quotations are submitted by filling out forms, rather than by composing letters. However, should you be required to submit a bid or quotation only occasionally, and you are not supplied with a standard form to execute, the following models will guide you in preparing them.

Bids and quotations are not the same thing. A quotation is the furnishing of prices for a client's information. The client may or may not place an order, and does not imply that an order will definitely be forthcoming as a result of furnishing quotations. A bid, on the other hand, is furnished in the expectation that someone, presumably the low bidder, will be awarded the sale. In business and industry, however, little distinction is made between the terms, and we will use the terms more or less interchangeably here.

The models used in this chapter are rather simple bids or quotations, with relatively small purchases at stake, usually, and often submitted to people the bidder knows quite well. Hence, they tend to be rather informal and straight to the point. Examples of rather simple bids by letter are shown in Figures 5.6 and 5.7.

Proposals

For many kinds of contracting, proposals are among the most important marketing tools, if not the most important. Contracting implies custom work. When the client is able to specify exactly and in detail what is to be done, it is practical to turn to sealed bids and award contracts to the lowest bidder. When it is not possible to specify in any great detail what is to be done—i.e., when the candidate must be asked to suggest a course of action or program to solve a problem that can be defined only in terms of its symptoms—the client is all but compelled to ask for proposals. The client must then judge which proposal is most acceptable and award the contract accordingly. Thus the importance of proposal writing skills: The quality and persuasiveness of the proposal is the determining factor in the award. In this kind of contracting arena, the successful contractor is one who writes the best proposals.

Proposals are of more than one kind, but there is an immediate division into two broad classes, the formal and the informal or letter proposal. This is primarily a function of the size of the project and, inevitably, the size of the proposal required. The project that will consume many weeks or even months of effort and cost many thousands of dollars will require a formal proposal of many pages, bound as a manual, carefully illustrated and produced at considerable cost to the

FIGURE 5.6 A Simple Bid by Letter

Dear Mr. Mandela:

Forthright Steel Rule Dies, Inc., is pleased to supply the following quotation, as you requested in your letter of last week:

1. For fabrication of cutting die, per the $ 378.00
   drawing supplied, including all labor and
   materials, using standard gauge steel
   cutting rule

2. For cutting 20,000 pieces, using stock   <u>$1,400.00</u>
   supplied by you

                            Total $1,778.00

Note: This quotation is based on standard work-week schedules (estimated time six weeks).

Thank you for inviting our quotation. It is a privilege to respond to your request, and we will be honored to handle this requirement for you.

Cordially,

Jean Granger
Standard Cutting Dies

---

proposer. The small project will justify only a small and informal proposal, written, usually only a few pages long and in general correspondence style. The letter proposal is a special, one-of-a-kind communication, addressing a specific problem or need of the respondent, and making some specific offer to satisfy the need or solve the problem.

    As an independent contractor, you will probably pursue projects and contracts of modest sizes and so will probably write letter proposals more often than you will formal proposals, although we will not entirely

FIGURE 5.7 A Detailed Quotation

Dear Ms. Wister:

The Four-Star Printing Company takes pleasure in responding to your invitation to bid for the printing and binding of 5,000 copies of your annual report, as specified in your request. As you requested, our bid includes a listing of the principal items and their cost, as follows:

1.	Negatives, color separations and plates	$3,756.00
2.	Printing, 3-color process, 60,000 impressions, including makereadies and press washes	$ 989.00
3.	Folding, binding and cutting	223.00
	Total	$4,968.00

Thank you for giving us the opportunity to offer our bid.

Sincerely,

Lars Olson
Vice President

---

neglect the latter, but will cover it here too. It is likely that you will find occasion to write a small formal proposal.

When Proposals Are Called For

Proposals result from clients' specific requests for proposals, as noted earlier, but they also result from the contractor's own initiative, usually following an initial meeting and discussion of a sale such as those listed here.

The size of the proposal is usually in some proportion to the size or complexity of the project to be undertaken. Obviously, the client who is going to make a large investment will want to study the pros and cons of each proposal submitted. Hence it follows that a letter proposal will be used, normally for relatively small or uncomplicated projects. At the same time, the project may call for providing all standard items, easily identifiable as commercially available, such as luncheonette equipment and fixtures, so that there is no complexity to the requirement, despite overall cost.

Why Clients Want Proposals

Each contractor and client has his or her own reasons for wanting the recommendations to be on paper, as a written proposal, rather than verbally or as a simple bid or quotation. A request for a bid or quotation requires the client to know precisely what he or she wants and to be able to specify the requirement in absolute detail. Frequently the client cannot do that or, even if he or she can, may prefer to benefit from the supplier's experience and ask for recommendations and costs on paper, subject to extended study, evaluation and negotiation. This is particularly true when the client asks several suppliers to submit proposals to be evaluated competitively.

Why You Should Want To Write a Proposal

If you have a good marketing instinct, you have your own good reason for preferring to supply a written proposal: A proposal, formal or informal, large or small, is a sales presentation. It enables you, as the seller, to argue your case for what you offer, and to argue that case almost at leisure. It overcomes that possible disadvantage in face-to-face selling of the impatient or overly busy client who does not permit you to make a full presentation of all your sales arguments. In a proposal you have the time to think out all your sales arguments and the opportunity to make your complete presentation without interruption. The client, too, may read and reread your presentation, studying it carefully, instead of relying on memory.

This means that you should welcome every opportunity to submit a proposal, even to the point of making a distinct effort to do so whenever you do not close the sale in a face-to-face meeting with the client. The smart marketer takes the initiative in situations where it normally requires two or more presentations and negotiations to consummate a sale, and advises the client that he or she will follow up promptly with a written proposal. No sales contact should ever be permitted to end without either an order or a definite next step planned and projected. For certain kinds of sales activity the proposal is the logical and best planned next step.

What Belongs in a Letter Proposal

Obviously, a letter proposal is going to present a sales argument. It generally considers one of two basic sales problems. In some cases you have to sell against typical resistance to change; that is, you must persuade the client to try something new and different. In other cases, where the client is definitely going to buy the service or product from someone, you must sell against competition, trying to persuade the client that your program or service is better than that offered by others. In practice, some sales situations are hybrids or combinations of both circumstances.

The letter proposal is normally contained primarily in a few pages— three to six, probably—that should offer the entire presentation. However, it may be accompanied by a brochure, catalog sheet, specification sheet or standard price list, where standard printed materials are appropriate. That would be the case where you propose a design which uses standard, off-the-shelf components, for example, or where you enclose printed materials to document your claims.

All the principles advocated for sales letters are equally appropriate here, with two basic differences:

1. The proposal is normally longer and more detailed than the mass-mailed form sales letter.
2. The proposal is based on knowing the client's specific need or problem and so may focus sharply on that and ignore other, irrelevant considerations.

The fact that a letter proposal is relatively brief and informal does not alter the fact that it is a sales presentation and must include the same kind of information that a multivolume, major proposal for a multimillion-dollar project would contain. The only differences lie in amount of detail and format considerations.

A client may call for proposals, issuing a general invitation to interested contractors to respond. But that is not always the case. Proposals in general are used in many kinds of sales or marketing situations, calling for a more formalized presentation than a simple sales letter or quotation. A proposal is required or, at least, a good idea when your quotation requires some explanation. Here are just two general categories of situations calling for proposals, and a few examples of each kind of situation:

1. To design something involving products, with the product(s) the main item and the service more or less incidental:
    - Design, layout and supply fixtures for a retail establishment or other facility
    - Recommend and supply a computer system.
    - Develop a special product, for example, a movie or automation device.
2. To propose a custom service, in which the service itself is the main item being supplied, with any products incidental:
    - Interior decorating
    - Renovation
    - Engineering
    - Landscaping
    - Training

A Suggested Format

Sometimes the client prescribes a format and requests information other than that listed here, although that is rarely the case with letter proposals, especially when the client has not requested the proposal. You may be quite informal, but you nevertheless must cover seven key points, preferably in the following order. Proposals, however, are custom presentations for custom work, and so can never follow rigid rules, but should be in accordance with whatever original understandings and discussions that preceded and led to the proposal.

1. Introduction: title, background of submittal and introduction to the proposer (you)
2. The client's perceived need or problem
3. Possible approaches and pros and cons of each
4. The selected approach and rationale for it
5. The specifics of what you will do, provide or both
6. Costs
7. Useful information about your own experience, qualifications, capabilities and references

Specific formats and content are up to you, generally, but those apparent in the models that follow are suggested, insofar as they fit your own situation.

A sample letter proposal offering a needed service in which you would also be supplying a product, with the service being almost incidental, as in the first category listed, is shown in Figure 5.8.

In other situations, the service may be far less significant than the product involved. As a computer consultant, for example, you might set up a computer you provided your client, but that would be a relatively minor service, although it would be a legitimate service for a client who could not or preferred not to do his or her own installation and set up.

Not all cases are so clear-cut, however, in the relative importance of the service to the product. In contracting to design and create a custom product for a client, the design work might or might not be a major element of the contracted obligation. The labor of designing a new computer program can and probably will be quite a significant part of the requirement. On the other hand, designing a newsletter for a client may very well involve less effort than the editorial work of preparing the copy, doing the layout and the other work of producing an actual issue of the newsletter.

The examples of the second category are therefore examples of contracts in which the product is relatively insignificant, as compared with the service required. Figure 5.9 illustrates this in that the author of the proposal felt the need to be quite detailed about anticipating, explaining and justifying the costs. Still, despite the detail required, the writer felt that she could present a convincing case in a letter proposal. Of course, she supplemented the proposal by enclosing some of her standard literature to demonstrate experience and capability, as she notes in her letter.

The subject of formal proposal is an important one for an independent contractor, and requires lengthy discussion, meriting its own chapter. The next chapter will discuss the subject at length.

FIGURE 5.8 A Sample Letter Proposal

Dear Mr. Warrington:

Kensington Lighting & Fixtures, Inc., is pleased to respond to your invitation to propose the provision and installation of the most suitable fluorescent lighting fixtures for your new warehouse addition. We have had the privilege of providing industrial lighting fixtures to many area firms over the past 27 years (see the appended brochure for details) and we are grateful for the opportunity to offer you our recommendations. We believe they will save you considerable expense, while giving you great satisfaction.

Before preparing this proposal, one of our most experienced representatives visited the location and studied your need as you described it and as it appeared to him, while considering most carefully the conditions under which the fixtures are to be used and the costs for various alternatives.

Inasmuch as the warehouse space is almost entirely clear span and is to serve for the storage of raw materials in bulk, we judged that you would have need for and should be interested only in purely industrial fixtures, providing good illumination without regard to cosmetic effect and the greater cost of the latter, so we confined our considerations to easy-to-install industrial fixtures, fixtures that will save you money in installation labor, as well as in their purchase costs.

The fixtures we selected and which we propose to furnish are extremely durable, quite easy to install and yet quite inexpensive, with a firm one-year guarantee. They are specified in the appendix to this proposal and illustrated in the manufacturer's brochures (enclosed). We sincerely believe that this choice represents the best value for your need, although we carry many other models in our regular line.

We judged the number of fixtures you needed by counting the number of terminations provided by your electrical contractor, whom, we presume, was working from approved architect's drawings. We found 32 such connection points provided, and for the amount of space represented by the new addition, we believe that 32 four-lamp fixtures (each lamp of 40 watts), with reflectors, will provide adequate

FIGURE 5.8 (continued)

illumination. On that basis, we propose to supply 32 such fixtures, and our cost estimate, included here as an appendix to this proposal, reflects that.

The estimate provided here includes complete installation and provision of the fixtures specified, fitted with lamps, starters and ballasts. Installation includes solderless connectors, suspension chains and all other accessory items required for normal installation.

These fixtures and installation are available for immediate delivery and service. A telephone call will bring our installation team to you, ready to begin work, within 24 hours.

I include here a catalog sheet, which provides a photograph of the fixtures proposed with all pertinent electrical and mechanical specifications.

Please feel free to call on me for more information or to discuss alternatives, should you wish to consider other types of lighting fixtures.

Thank you for the opportunity to furnish this proposal. I look forward to being of service to you.

Sincerely,

Wilton James
James Electrical Service

FIGURE 5.9 A Detailed Letter Proposal

Dear Mr. Hannigan:

As I promised, when we discussed your problem at the recent Education and Training convention last weekend, I did some research immediately upon getting back to my office. I found that we did, indeed, handle a quite similar problem several years ago for the Williamson Electrical Parts Company. They, too, were suffering the drawbacks of an overly rapid expansion—a sharp drop in productivity, as a result of breaking in large numbers of new employees at the same time, contributed to in no small measure also by the tremendous burden this placed on their supervisors. (Some enlightening details of that project are presented in an appended brochure, along with other information about MW Designs, Inc.)

Williamson retained us to take over the training from their overworked supervisors, so they could concentrate on maintaining productivity with the older workers.

It worked out well, for Williamson was thus able to absorb the expansion much more rapidly and they got back to full productivity in about six months. (We worked closely with John Murphy, their Director of Administration, and I am sure that Mr. Murphy would be quite willing to talk to you and confirm what I say here.)

What is needed, Mr. Hannigan, is a three-phase program. The first phase would be a formal Task Analysis, to do the basic research, establish the specific parameters of the problem, and draw up the specifications for the phase-two training program. The second phase would cover the actual development of the training program and administrative plans to conduct it. And the third, final phase would be to implement the plans by actually conducting the training.

All of this would have to be done in close coordination with your staff, of course, and all plans would be subject to review and approval by whomever you appoint to oversee the program for your company.

FIGURE 5.9 (continued)

The schedule of main events would be approximately as follows:

1. Task analysis and training-design development	30 days
2. Training program and materials development	60 days
3. Training implementation	90 days
Total	180 days

Costs are estimated as follows:

Task analysis:	$11,500
Training development:	33,700
Training (our staff):	27,500
Total	$72,700

We are prepared to get under way with this important program on a week's notice, and I am prepared to quote the above schedules and costs as firm if we consummate an agreement within the next 30 days.

Of course, I welcome any questions you may have and will be pleased to meet with you for discussion, to make a formal presentation to you, with or without your staff and to enter into serious negotiations immediately.

Under the circumstances I think you will agree that there is nothing to be gained by waiting, and delay is likely to be, in fact, quite costly.

Thank you for the opportunity to offer this proposal. I will be most pleased to serve your fine company in this matter.

Most cordially,

Marie Wilson
MW Systems, Inc.

The Formal Proposal

The formal proposal is the best opportunity you get to sell yourself and your services. It is important to take the fullest possible advantage of it.

The Proposal Concept

The general philosophy and rationale for a formal proposal is the same as that for a letter proposal: Introduce yourself as the proposer and explain the reason for and circumstances of your submittal; how you perceive the need or problem; your views and discussion of possible approaches and pros and cons of each; your choice of approach and rationale of your choice; the specifics of what you will do or provide and when; information about your own experience, qualifications, capabilities and references; and costs. In many cases, such as government contracting, costs are required to be in a separate document so that technical evaluation may be made objectively and without regard to costs.

Major Elements and Structure

Formal proposals are the same as informal or letter proposals, in principle. The differences are in length, format and, of course, depth and scope of details included. Where informal proposals are in letter formats, formal proposals follow the format and organization principles of manuals and books, bound within covers. Following is a list of the items that generally are found in a formal proposal, with general information describing or further identifying each item, as appropriate. These are summary descriptions, and although I recommend the order of presentation shown here, in many cases the client has mandated a format he or she requires you to follow or where the format advocated here is inadequate for other reasons. In both cases, your proposal format must be expanded or modified accordingly. The following items and summary descriptions or definitions are therefore offered as preliminary and general guides to formal proposal format.

Cover

The cover should include the title of your proposal, descriptive or definitive of requirement or program proposed, with pertinent references, if any (e.g., identification of request for proposal [RFP] or tender to which proposal is responding), name and address of proposer and date. (The date is optional, and is not always used on cover.)

Front Matter

The formal proposal includes front matter, as a book or manual would. Following are the principal kinds of front matter appearing in proposals.

- Title page: This includes much the same information as the cover, plus the date and often a proprietary notice covering the information therein.
- Abstract: This usually is called "executive summary" and summarizes, highlights and dramatizes major selling points.
- Table of contents: This is a list of the chapters, and often includes major subheadings under chapter title listed and a list of figures and tables.
- Preface or foreword: An optional item, this may be used as desired, although its purpose is often accomplished by the executive summary, letter of transmittal or both.
- Response matrix: This is a very useful device when the RFP and the requirements are multifaceted and somewhat complex. It's a matrix of RFP requirements and proposal responses, guiding the reader to all pertinent information mandated by the client and responded to in the proposal. Its purpose is to ensure that the

client easily can verify that you have responded to all requirements, and to ease the client's evaluation of your proposal.

First Chapter

The objectives of the first chapter are as follows:

- To establish your identity and basic credentials
- To demonstrate adequate understanding of the requirement

This chapter contains brief introductory material, identifying proposer and requirement to which proposal responds, previewing what is to come in the proposal and summarizing the requirement in its essence. (Save the details for next chapter.)

Second Chapter

The objectives of the second chapter are as follows:

- To present detailed discussion or analysis of requirement
- To establish and sell your understanding of, approach to and capabilities to do the job (demonstrate your mastery of the problem)
- To prepare the groundwork for next chapter

This chapter contains an extended discussion of requirement(s), identifying your approach (in principle) and explaining and arguing the merits of your approach.

Third Chapter

The objectives of the third chapter are as follows:

- To describe and stipulate your proposed program, showing your practical implementation of the approach you identified earlier
- To list the specifics of what you propose to do and provide
- To demonstrate that you can carry out the plan outlined by showing operating methods
- To present your schedule

This chapter contains the specific proposed program or project in detail: what, when, where, how, why, by whom and so on.

Fourth Chapter

The objectives of the fourth chapter are as follows:

- To document your resources—facilities, staff, experience and record of accomplishment
- To prove that you are a reliable contractor, as well as a capable one

This chapter contains qualifications of proposer, descriptions, lists of facilities, resources, experience, references and all supporting details.

Appendixes

The objectives of the appendixes are as follows:

- To supply supplemental information of interest to some, but not necessarily all, your readers
- To document and validate statements made in proposal

The appendixes contain additional information, if and as necessary to present material that should be appended, rather than included in body of proposal, such as copies of papers, drawings, photographs, reprints, backup statistics, slides and other such material.

The models offered to exemplify and illustrate these ideas will necessarily be brief and partial ones, greatly condensed for the obvious reason that even a single formal proposal is generally book-size and could not be accommodated in the space available here. However, the models will illustrate the main principles presented.

The elements of a typical formal proposal have been listed here in the order of their usual appearance to define the sequence and recommended format. In the following discussion, however, the sequence will be somewhat different, beginning with a brief discussion and samples of titles and then going to discussions and models of the text.

General Observations

Many of the points about to be made are minor in themselves, but taken altogether they can have a major impact and make a great difference in the success of the proposal.

Proposal Titles

Many people try to create short and "catchy" titles for books and headlines for advertisements. Regardless of any merits in that idea, it is entirely unnecessary and even unwise to do this when titling a proposal. It is far more important to use a title that states clearly what the proposal is about so that there is no confusion about it. For one thing, some clients issue many RFPs and may have more than one set of proposals arriving for evaluation. They therefore need to distinguish your proposal from others without difficulty. Keep the title as short as possible but long enough to identify it clearly and easily by virtue of the text of the title, without the need to rely on reference to the client's request number (although that should also be furnished, when known). Here, for example, are a few typical proposal titles:

- A Proposal To Research Tobacco Consumption in the Middle Atlantic States
- A Proposed New Method for High-Speed Inter-Computer Communications
- A Proposal To Design and Develop a Standard Navy Teleprinter
- A Proposal To Develop and Deliver a Training Program in Blueprint Reading
- A Proposal To Design and Develop a Management Information System for the Wendex Corporate Headquarters Central Computer

Format

Although a most general format and organization have been suggested here, a great many variations are possible. Some of these are of negligible importance and have no bearing on the outcome of the competition, while others are quite important and may very well affect the outcome. This is an appropriate time to review several of these matters briefly.

Type Spacing Most proposals are typed or printed today via a word processor, which is essentially the same thing as typing as far as physical appearance is concerned. There appears to be a school of thought that double-spaced copy is easier to read than single-spaced copy, and so double-spacing is quite common in proposal preparation.

There is not a great body of evidence on this matter of type spacing versus readability, but there is another school of thought that inasmuch as we all learn to read single-spaced printed material, we are conditioned to and are probably more comfortable with typing that is single-spaced. At the same time, double-spaced copy is somewhat less formidable in its appearance, although it is doubtful that that is a factor of any great importance. My own preference and my own recommendation is to use single-spaced copy or, as a second-choice alternative, one-and-one-half spaces between lines, an option available on most modern typewriters and printers. (Many printers offer even more options in spacing.) It does help readability, however, to keep paragraphs as short as possible—i.e., to start new paragraphs as often as possible, and to have a clear topic sentence introducing each new paragraph. That greatly reduces the required attention span and helps the reader focus on the subject of the paragraph.

Type Sizes Typewriters have traditionally offered two type sizes, elite or 10 point, and pica or 12 point. Today, under the influence of modern typewriters and word processing computers, these have become known as 12 pitch (12 characters to the inch) and 10 pitch (10 characters to the inch), respectively. This can be somewhat confusing, because 10-point type equals 12-pitch type and 12-point type equals 10-pitch type. Obviously, 12-point/10-pitch (pica) type is the larger size and thus perhaps is easier to read. Ordinarily, both are acceptable, but smaller type sizes than 10-point (elite or 12 pitch) are to be avoided.

Chapter Titles, Heads and Captions Almost all proposal writers who have not been specifically trained in writing persuasive copy use unimaginative generic titles, heads and captions that contribute nothing to the effectiveness of the proposal. Most will title the first chapter "Introduction", which appears reasonable enough, since that is the purpose of that first chapter, but it contributes little or nothing to sales persuasiveness. A properly chosen chapter title, however, may contribute substantially to the success of the proposal by commanding special attention, dramatizing or highlighting some key point or summarizing an important sales argument. For example, a small firm in pursuit of a contract with a government agency learned that the agency was not pleased with the performance of the incumbent contractor and would welcome a proposal that would justify replacing the incumbent with a new contractor. Accordingly, the firm titled the introductory first chapter of their proposal "A New Broom," making a clear point, of course. The idea, overall, is to make every title of any kind do two things: deliver a message and sell.

The same philosophy should be applied to the selection of words for all heads and captions. Again, it is desirable to make the heads and captions short, but only when it can be done without sacrificing the message and its impact. Actually, you should write the original draft to say it all, and then study ways to make it shorter and "punchier" without losing anything. For example, instead of "Proposed Schedule," use something such as "A Schedule Responsive to the Need" or "A Time- and Money-Saving Schedule" as a headline to introduce a discussion of scheduling or as a caption for the schedule itself. (Of course, you must justify what the headline or caption says, and it should address an objective you have good reason to believe is important to the client.)

Never include a title, head, subhead or caption that does not make some direct and positive contribution to the sales effort represented by the proposal. Each is an opportunity to sell what you wish to sell—or a lost opportunity.

Models of the Chapters

The examples used in this section are based on work to be performed by a staff of several people. You must adapt them to your own needs, of course, in the event that you will do all the work alone, although you may wish to utilize the services of associates. If you alone will do the work, treat your company as a separate entity and yourself as its chief employee.

As already noted, space limitations make it necessary to present only abbreviated models of the several proposal chapters, virtual abstracts of such chapters. By the same reasoning, space limitations prevent the inclusion of illustrations referred to in some of the models. In fact, illustrations, conceived and used judiciously, greatly reduce the amount of text required.

The Introductory Chapter

The first chapter should be relatively short, in relation to the overall size of the proposal, its number of pages consisting of probably not more than two or three percent of the total number of pages. In many cases, it can be as brief as one page. Two examples of such a chapter are offered in Figure 6.1.

Note one highly significant thing about these two models of a first or introductory proposal chapter: Each highlights a specific problem or important consideration of the requirement and promises a solution, without actually revealing or defining that solution in anything approaching detailed disclosure. Instead, both promise a solution to be revealed and explained in the next chapter.

That method of presentation and introduction accomplishes several important goals:

- It creates the necessary transition from the first to the second chapter. A transition or "bridge" should always be present between all narrative elements—sentences, paragraphs and chapters—to continue the train of thought and build the chain of logic steadily.
- Even more important than that, however, is the creation and maintenance of interest. There is an element of suspense in telling some but not all of an important matter, thus motivating the reader to hurry on to the next chapter to see what the promised outcome will be. It is also an excellent competitive strategy. In fact, this technique actually dramatizes the message and thus greatly magnifies the impact of the ideas offered.
- Finally, there is one more benefit: This enables the next chapter to "start off with a bang"—to have a built-in interest factor—adding to its own effect. The second chapter should take up immediately where the first chapter ended, discussing the client's problem or requirement.

Second Chapter

You must persuade the client to approve and accept many things in any proposal, but these are the three main "sales" your words must make:

1. Sell your plan or approach to solving the client's problem and satisfying the requirement.
2. Establish your capability for carrying out the plan successfully.

FIGURE 6.1 Two Examples of a First Chapter

CHAPTER I

A Response That Matches the Need

A Well-Qualified Offeror

Hackney & Williams Software, Inc. (HWSI), is pleased to offer Martinside Manufacturing, Inc., our services to design and develop a management information system (MIS) to be used throughout the Martinside organization, as specified in Martinside RFP M-87-0032. HWSI has been active in software development for 14 years and numbers many major corporations and government agencies among our clients, many of whom retained us to develop the same type of program as that discussed in these pages.

Details of our experience, résumés of proposed staff, client lists and references will be found later in this proposal.

A Special Insight into the Requirement

The RFP describes symptoms, stating that your internal computer reports are neither timely nor otherwise helpful in detecting and correcting problems in time to forestall disastrous consequences. It speculates that an improved or perhaps new and different computer (MIS) program is required, acknowledging that this will require first a complete analysis. We believe that this is a premature conclusion, since there are several other possible causes, such as poorly designed data-collection forms or weaknesses in data-collection methods, for the production of reports that do not meet the needs of Martinside. That is, the RFP and its work statement describe symptoms, but not necessarily causes or problems *per se,* and it is important that we not confuse the two. We believe, therefore, that the initial investigation to confirm the preliminary diagnosis must consider these possibilities, and we will explore and discuss these in depth in the next chapter.

FIGURE 6.1 Two Examples of a First Chapter (continued)

CHAPTER I

A Total, One-Stop Capability

Over a Half-Century of Experience in Your Service

Accurate Litho, Inc., takes pleasure in responding to your request for a proposal offering our services to produce the 1987 Magnetic Media Corporation Annual Report. Accurate Litho has been producing such annual reports and other fine, process-color literature for leading corporations in all industries since 1911. We are completely equipped with in-house staff specialists and facilities for typesetting, illustrating, layouts and all other functions required to take the job from edited and approved manuscript to printed and bound copies delivered to whatever destination you specify on a guaranteed schedule.

That is a pledge, and details of our facilities and resources, presented later in this proposal, will demonstrate that this is not an idle pledge.

The Special Problem We Must Solve

We have studied your statement of need carefully, and we recognize a problem that may not have been apparent to you when you issued this request. That is the problem of a schedule that will be difficult to meet: You will not have final copy ready until the 15th of next month, but require the printed and bound product by the first day of the following month, although we will have to begin with raw manuscript to be typeset, proofread and corrected before we can even fit copy and make layouts.

We are well aware that it has become something of a bitter joke that printers always promise delivery and rarely meet the schedule. That is not the case here. We are prepared to guarantee delivery on schedule, and in the next chapter of this proposal we will explain how we propose to handle this project so that we can make that guarantee and live up to it.

3. Prove your dependability as a contractor or supplier.

Although all chapters should be written with these three main factors much in mind, the first item should be the chief focus of the second chapter. The second chapter begins with a full discussion of the requirement and goes on to explain your approach to satisfying it and presenting your rationale. When you have finished this chapter, you should have prepared the client for the next chapter. This chapter should end with a transition that telegraphs or at least foreshadows the details that will follow in the next chapter. This chapter is thus a technical discussion, normally elaborating on your understanding of the client's requirement and explaining your rationale for the approach you propose, while you sell yourself in the process.

Examples of the second chapters can be found in Figure 6.2.

At least three major points should be made with respect to the second chapter examples. Despite their brevity—they likely would be many times longer than these brief samples—they illustrate the main principles, and especially the following:

- They take up exactly where the previous chapter ended. The transitions from one to the other are very much in evidence.
- They prepare the reader to go on to the next chapter, with clues of what is to follow—transitions between chapters once again.
- Each is based on a strategy that claims and demonstrates special attributes for what is proposed, while suggesting at least by implication that all competitive proposals ought to be examined to see if their authors have perceived the obvious problems and provided some practical means for coping with them successfully.

The Third Chapter

The third chapter, at least as defined in the format and organization recommended here (see Figures 6.3), is the chapter that usually is incorporated specifically into the contract itself, directly in the language, as an exhibit or by reference. That is because it includes the specific details of what you pledge to do or deliver to the client, or both, along with when—schedules, that is. It is, in fact, the proposal *per se*, where the other elements of the proposal are really ancillary to it, supporting it directly and indirectly.

For that reason, the third chapter in the format recommended here assumes its own special importance. Where the main importance of the

FIGURE 6.2 Two Examples of a Second Chapter

CHAPTER II

A Frank Discussion of What Is in Your Own Best Interests

From Experience Comes Wisdom

Our experience in developing new management information systems (MISs), as well as troubleshooting, redesigning and reprogramming existing ones, has taught us certain truths:

1. The possibilities for system malfunctions—that is, the number of elements that are vulnerable in most systems and so may be responsible for problems—are so numerous that it is most risky to take anything for granted. Every possibility must be considered and investigated before drawing conclusions.

2. It is as true as ever that the output of any system, no matter how well designed, cannot be any better than the input. Therefore, as much attention must be devoted to the data-collection forms and procedures as to the system and programs overall.

3. It is often quite easy to confuse the symptoms with the problem—to identify a symptom as the problem, that is—and so pursue the wrong objective.

4. As a consequence of these factors, it is almost always necessary—or, at least, certainly the wisest procedure—to begin each major project with a front-end analysis to ensure that we have a firm identification and definition of the problem so that we can establish our objectives accurately before investing major effort.

5. A proper problem definition is itself almost invariably an arrow pointing directly to the possible solutions or, at least, to the proper approaches to solutions.

It is, of course, very much in your own interest that we perform this analysis, and in fairness to you, as well as to our own reputation, we could not approach this project without making this clear to you.

FIGURE 6.2 (continued)

Identifying the Most Economical Solution

Your work statement is based on an assumption that a new and more efficient MIS must be designed and programmed. We have no quarrel with this, of course, should it prove necessary to pursue this course. However, even in those cases where we have found faults in the MIS to be the main cause of management-reporting problems, it has not always proved to be necessary or most efficient to develop a new system. Often it has proved entirely feasible to modify (reprogram) it, salvaging much of the original work.

The advantage of this is an obvious savings in money, but there is also a savings in time, which is often an even more important consideration.

A Practical Approach

In consideration of these factors, HWSI proposes a two-phase approach, per the following:

> Phase 1: We will conduct an initial analysis, including a 60-day study of the existing system, forms and procedures, followed by a definitive report to you. The report will include all information necessary for a second, follow-up phase.

> Phase 2: A project, based on and defined in the report on the phase 1 study, along with a proposal listing time and cost estimates for the second phase.

In this proposal, therefore, we will estimate the procedures and costs for only the phase-1 analysis, with specific details to be presented in the next chapter, and projecting possible phase-2 events in only the most general of terms.

FIGURE 6.2 Two Examples of a Second Chapter (continued)

CHAPTER II

Meeting an Impossible Schedule Successfully

Our Guarantee To Meet the Date

Fifteen days is a difficult schedule to produce printed and bound literature starting with manuscript and glossy photos, which must appear as in-text illustrations. Many steps are required, including the following major ones:

1. Set type and print galleys.
2. Proof and correct galleys.
3. Make type changes and print page proofs.
4. Make negatives, including color separations.
5. Make plates.
6. Print reports.
7. Bind and trim.

Other steps are required, however, which must be carried out in order for these to be done. Page layouts, as roughs and again as comprehensives must be made up, and finally mechanicals—final copy from which negatives and plates can be made—must be made up. And all of this must be carefully coordinated to meet even a normal schedule (which would ordinarily be six weeks for this task).

In most cases, to do all this in 15 days would be utterly impossible. Even for our own firm, which is accustomed to, is equipped for and specializes in "quick-response" projects, this task required a great deal of careful planning before we felt able to offer this proposal and pledge—actually, guarantee—to meet the schedule. Even given the special advantages we enjoy in having most of the necessary skills and personnel available in-house at Accurate Litho, and given the special conditions we shall cite as essential to meeting the schedule date, it will be a most difficult task. It will be accomplished successfully

---

FIGURE 6.2 (continued)

---

only by the greatest of diligence and resourcefulness, which, fortunately, we are equipped and prepared to provide.

Essentially, the special conditions require your own spontaneous reviews and approvals—permission to proceed—as we proceed through the steps. That means that we will ask you to have someone with authority to approve copy come to our facility to carry out those reviews and give approval swiftly, as we go. Given that, we can guarantee success.

In the next chapter, where we present the absolute specifics of what we propose to do to meet the required delivery date, we present the total set of necessary conditions. They do not demand a great deal of Magnetic Media Corporation, however, other than on-time and as-required attendance at our plant by someone with authority to make swift reviews and give permission to go ahead so we can adhere to what will be a very tight regimen.

---

second chapter is that it must sell the client on the approach and plan, the main importance of the third chapter is that it presents the specifics of what you propose to contract for—what you are to do or deliver. You will be bound by what is in this chapter. Conversely, make what you are pledging to do or deliver crystal-clear here to protect yourself in the event of a dispute later over what you have agreed to. That is an essential in proposal writing; carelessness here can be costly later.

The format shown in Figure 6.3 calls for the résumés of principal staff to be included at the end of the third chapter, since that is the chapter that is dedicated to providing the specific details. That may be your own résumé alone, unless you are planning to use associates, employees or subcontractors.

Remember, again, that these examples are very much abbreviated samples of the typical formal proposal, and are used to illustrate main principles only. In the first project, which details professional, highly specialized and custom services, the staff and organization are of primary importance, and so the stress is made there. In the printing proposal, however, the competence of the staff to carry out what are

FIGURE 6.3 Two Examples of a Third Chapter

CHAPTER III

The Ingredients of Success

A Dedicated Project

This is to be a special project at HWSI. That means that a staff will be assigned directly to this project, with these project assignments their first priority, taking precedence over all other work. We can thus guarantee a continuity of effort and purpose: there will be absolutely no diversions of staff or effort from this important project.

A Highly Experienced Staff

For the first-phase work, HWSI will assign two top scientists to the project.

The project director and principal investigator will be Dr. William Langston, a vice president of HWSI and a veteran of more than 20 years in this field, specializing in management sciences and management information systems in industry, business and government. He brings to this project the benefits of a distinguished academic background, as well as extensive and relevant experience in both government and private industry, where he directed a number of major MIS program developments.

Dr. Langston will be supported by Terry Willets, serving as second investigator. Mr. Willets, who earned his MBA in management sciences at Wynnefield Polytech, has been with HWSI for nearly 15 years and has worked on and led many MIS design programs at HWSI and elsewhere in his career. In fact, his own impressive experience and accomplishments rival those of Dr. Langston.

Fully qualified support staff will also be provided, and detailed résumés of all principal staff members will be found later in this chapter.

FIGURE 6.3 (continued)

The Deliverable Item

We estimate that the report of the first-phase study will be approximately 75 pages long, including at least six flowcharts depicting the preliminary design proposed for second-phase development, as part of the final section of the report, Recommendations. (That may represent a redesign of the existing system, or it may represent a proposed new design, depending on the results of the study, for which full rationales will be provided.)

The report will make a full disclosure of findings, with detailed recommendations for the second-phase follow-up. These will include a critique of the present system and of its forms and procedures, with suggestions for improvement or replacement wherever and of whatever deemed necessary, and full explanations to justify all recommendations.

The report will include the information listed below and follow this general format:

- Problem definition and statement of task objective
- Research and data-gathering methodology
- Narrative account of data gathering
- Presentation of raw data
- Account of data reduction methodology
- Presentation of reduced data
- Analysis and rationale
- Conclusions
- Recommendations

CHAPTER III

The Program To Meet an "Impossible" Schedule

A Plan for Effective Action

To understand and appreciate the details of our program for meeting the extraordinarily difficult schedule requires an understanding of

FIGURE 6.3 Two Examples of a Third Chapter (continued)

the normal production tasks and time involved for each, as exemplified here. (These are the six main tasks required for the program called for by this RFP, many of which are not really printing and binding tasks, but are editorial functions that Accurate Litho offers as a convenience to our patrons.)

1. Set type, print galleys	4 weeks
2. Proof and correct galleys	3 weeks
3. Make type changes, print page proofs	2 weeks
4. Make negatives, color separations	6 weeks
5. Make plates	1 week
6. Print, bind and trim	1 week
Total	17 weeks

Even an accelerated schedule would ordinarily mean trimming this schedule by only one-third to one-half the allotted time. To do this entire job in 15 days—only two weeks, that is—means trimming the normal schedule by nearly 90 percent!

These tasks are normally carried out sequentially in serial fashion, and in fact in some cases there is no alternative to serial sequencing. However, it is possible to perform some of these functions or portions of these functions concurrently—e.g., begin the next function as an overlap—to shorten total elapsed time for the project.

Other measures are possible, too. Printing on the calendered stock called for presents a drying problem, especially with multicolor printing, so we will have to accelerate the drying process by using a special infrared drying line available in our plant.

Some of these measures mean undertaking some risk of redundant and inefficient efforts, but our plan minimizes that hazard, and Accurate Litho is willing to underwrite the risk.

FIGURE 6.3 (continued)

The schedule presented in Figure III-1, a milestone chart, and Table III-1, a matrix of events, illustrates our proposed method for accomplishing this near-miracle. These illustrate the process of concurrent and overlapping tasks, that is, and the sharply abbreviated schedules for each.

What Accurate Litho Pledges To Deliver

The end-item, as identified in the RFP and confirmed here, is a 24-page annual report, printed in black plus three other colors, with eight four-color photographs (halftones), printed in quantity of 12,000 copies, bound and trimmed, trim size 9 × 12 inches, delivered 15 days after final copy and photographs are received. (Details of stock, binding and other such details are included, listed on one of our standard estimating forms enclosed here.)

Accurate Litho will provide experienced and capable staff (see résumés appearing at the end of this chapter) to manage this program and perform all production tasks, per the schedule, as listed in this chapter. Magnetic Media Corporation will provide its own personnel, at the Accurate Litho plant, to review and approve copy and give prompt permission to proceed.

A Timely Schedule

Figure III-1 depicts the full schedule of events in the form of a milestone chart, which reveals exactly where, how and what efforts are to be expended, and when each task is to start and to end. The milestone-chart format enables you to see how the several tasks relate to each other in their phase and functional relationships.

Résumés of Key Staff

The résumés of the principal staff to be assigned to this project are offered here, and their assignments to the project are guaranteed. In the event that some contingency impossible to foresee now interferes with that plan, HWSI guarantees to offer other staff résumés for your approval before making other staff assignments.

usually routine functions is ordinarily taken for granted by the customer, so the emphasis is on the problem itself. In other words, the strategy in the first case was to prove the unusually great staff experience and competence, as demonstrated by résumés and in-text observations. In the second case the strategy is to "prove" that probably no one else is likely to be able to deliver on time by demonstrating the virtual impossibility of the schedule requirement, and then presenting, with apparent confidence, a plan for getting the job done on schedule.

The résumé model of the first case is built around a design to enable you to tailor each résumé to the proposal's specific requirements, and yet not have to completely rewrite each résumé for each proposal. The items that normally have to be rewritten are the "proposed assignment" and the "summary record of professional achievement and special qualifications." Usually the remainder of the résumé may be used as a boilerplate.

Of course, if you use a word processor, the job of tailoring those résumés to each set of proposal requirements becomes infinitely easier and more convenient. That is true, moreover, for the overall proposal effort, especially for the fourth proposal chapter in the recommended format.

The Fourth Chapter

The main objective of the proposal's fourth chapter (see Figure 6.4) is to demonstrate and build customer confidence in your reliability as a contractor, while also reinforcing and reaffirming your earlier evidence of your capabilities, as a secondary objective. Therefore, the main thrust of this chapter is presenting a record of your achievements—current and past projects of a relevant nature, along with identification of customers; special credentials, such as letters of commendation, special awards, patents and other items; and details of resources—such as physical facilities, personnel, corporate organization and anything else that contributes to this objective.

Again, the examples are brief, and only typify the more extensive coverage that would normally appear in most formal proposals.

The fourth chapter presents your qualifications and resources to prove your capability, and whatever other evidence you can furnish, such as descriptions of past projects and names of references to verify what you say. It is a virtual company résumé, rather than your individual résumé.

FIGURE 6.4 Two Examples of a Fourth Chapter

CHAPTER IV

HWSI's Outstanding Record of Achievement

First, the Right Experience for the Job

It is no exaggeration to say that our unusually appropriate qualifications for the proposed project represent distinct assets for the Martinside Manufacturing organization, since these HWSI resources will be committed to success on behalf of Martinside. Among these assets is the particularly appropriate experience of Hackney & Williams Software, Inc., as demonstrated by the following few summaries of recent HWSI projects:

A Large Government MIS We have just completed final testing and debugging of a large management information system for the National Bureau of Standards (NBS) and demonstrated its operation to NBS, which has now accepted it. It was an unusually broad-based system for a large mainframe computer, representing a government investment of several million dollars and an eight-month effort. We are pleased to report that we delivered the product nearly two percent under budget and almost 30 days earlier than the stipulated deadline.

Contact to verify this: John Muller, Contracting Officer, (202) 555-1333

A Special Problem Solved for a Customer Modern Alchemy, Inc., a large midwest manufacturer of plastics and other products, has become a conglomerate in the chemical industries by acquiring 22 smaller companies over the past few years. Many of these companies had their own mainframe or microcomputers and management information systems, some of the systems custom-made for them, others commercial off-the-shelf programs. This polyglot assemblage made central management and reporting rather difficult for the corporate staff.

HWSI was asked to create a central MIS that would retain as many existing systems as possible, within the constraints that they had to be made compatible with the new, central MIS, and had to furnish the

FIGURE 6.4 Two Examples of a Fourth Chapter (continued)

information that Modern Alchemy corporate staff required and specified in their RFP.'

HWSI managed to create the new system and achieve the objectives by retaining and fitting into the new design nearly 90 percent of the existing systems, modifying only four of the older systems to make them compatible, and retaining the rest without change.

This was accomplished by HWSI within the budget and schedule originally proposed by HWSI.

Contact to verify this: Harry Mulligan, Executive Vice President, (312) 555-6667.

The Right Physical Resources

Hackney & Williams Software, Inc., occupies a modern six-story building of our own, in which we have over 47,000 square feet of office and laboratory space and 560 employees, including the following:

- Computer engineers and scientists
- Systems analysts
- Research specialists
- Programmers
- Library scientists
- Technical writers and indexers
- Support personnel

Our laboratory facilities include a mainframe computer and a minicomputer, with appropriate peripheral equipment (tape servos, etc), and more than a dozen desktop computers.

Figure IV-1 illustrates the corporate organization and shows where the proposed project will be established in the structure. Note that it is assigned a position that ensures it a high degree of visibility by top corporate management, a reflection of the importance we assign to this project.

FIGURE 6.4 (continued)

Figures IV-2 through IV-6 are copies of unsolicited letters of commendation from customers who chose to thus express their appreciation for the way in which we conducted their projects and satisfied their needs.

CHAPTER IV

Accurate Litho's Unusually Complete Facility and Organization

A Total Publications Organization and Physical Plant

Few printers maintain as complete an organization and set of physical facilities as those of Accurate Litho. For one thing, few printing plants have their own facilities for making the color negatives and separations for process color printing; most vend this part of the work to specialists. And few have a complete graphics arts and editorial staff on-premises full time, ready to support all publications requirements.

Within our modern three-story building is a complete printing and binding plant on the ground floor. This plant occupies 8,400 square feet on one level, with large web presses and small sheet-fed presses, a 36-inch camera and dark room and a complete bindery. The second and third floors, constituting another 14,600 square feet, house 234 employees in business offices and editorial offices.

Among the many similar projects we have handled in the recent past were these:

> 125,000 copies of American Tool Steel Corporation's Annual Report, similar to that referred to here, delivered four weeks after arrival of camera-ready text and photographs. Contact at American Tool Steel: Tony Morgan, (206) 555-9990.

> 3,000 copies of 456-page manual, for United Technical Services Corporation, including 38 process color photos, sewn bindings and paper covers, produced in six weeks. Contact: Barry Greenbaum, (415) 555-8677.

FIGURE 6.4 Two Examples of a Fourth Chapter (continued)

> Figure IV-1 is an organization chart, showing the entire Accurate Litho corporate structure. Figure IV-2 is a typical workflow diagram, reflecting our normal processes. Figures IV-3 and IV-4 are certificates of commendation from the U.S. Navy for an earlier project in which we met difficult deadlines.

Front Matter

The two items of front matter that need special discussion here are the executive summary and the response matrix. Ordinarily, the use of an executive summary obviates any need for a foreword, preface or abstract in the front matter. The response matrix is the other important item requiring special discussion, although brief mention will be made also of the letter of transmittal used normally in submitting a formal proposal.

The Executive Summary The executive summary is nominally supplied for the benefit of those executives in the customer's organization who would normally not, for whatever reason, plod through the entire proposal. This implies that the executive summary is an abstract of the proposal. The shrewd proposal writer will utilize the executive summary to summarize the most persuasive highlights of the proposal—i.e., the major benefits and other arguments for the selection of the proposer as the awardee.

The typical executive summary found in most proposals (see Figure 6.5), therefore, is a crisp, sometimes even staccato, recital of major arguments, in brisk telegraphic style designed for impact style as well as content.

The Response Matrix Unfortunately, many customers miss important points in reading proposals, and so fail to credit the proposal writer with having supplied adequate information or responding to the RFP request for specific data. In many cases this alone can cost you a contract you should have won.

FIGURE 6.5 Two Examples of Executive Summaries

Executive Summary

A contract award to Hackney & Williams Software, Inc., (HWSI), represent many benefits and advantages to Martinside Manufacturing, Inc., such as the following examples:

- Fourteen years of highly specialized and totally relevant MIS experience
- An insight that anticipates possible false starts and therefore prevents them through preventative measures
- One of the nation's leading experts in management sciences and systems leading the project, supported by another expert of almost equal credentials
- HWSI's perfect record of on-time and within-budget performance in 374 prior projects
- Program-installation backup support (guaranteed) should any problems surface
- Proximity of contractor's corporate offices

Executive Summary

Accurate Litho offers Magnetic Media Corporation many exclusive advantages in this proposal, owing to Accurate Litho's facilities and capabilities. The following are points which, we believe, must be considered carefully and thoughtfully in selecting an awardee for this contract:

- Accurate Litho offers a highly unusual—probably unique—set of publications capabilities under one roof.
- Owing to this remarkably comprehensive and complete set of capabilities and the unusual schedule requirement of the RFP, it is likely that Accurate Litho is the only contractor who can deliver this project on schedule.
- Accurate Litho offers a concrete, detailed and specific plan to meet the schedule.
- Accurate Litho guarantees to meet the schedule under the conditions set forth.
- Magnetic Media is justified in having faith in Accurate Litho's promise, for reasons set forth in this proposal.

Thus, one way to make the earlier examples even more effective is to follow each statement with a notation of where in the proposal the reader may find details of what the item claims. That leads us to the response matrix, which does that and more, although it is helpful to also note the chapters, pages or paragraphs that offer the details of the items.

The response matrix offers you a convenient and effective way to solve, even avoid, this problem. Simply put, the response matrix is a table that lists all the items the customer has requested in the RFP and specifies where in the proposal the requested information is supplied. That is, a response matrix is designed to help the reader verify the total responsiveness of your proposal and so maximize your technical score where there is systematic scoring, and to maximize the effect of your technical proposal in all cases. In general, it is a table in which are listed the requirements specified in the RFP.

The procedure is to make a list of all requirements listed in the request for proposals, and then prepare a simple table or matrix that lists these and identifies where in your proposal you respond to and satisfy each such requirement. It is usually advisable also to introduce this matrix and explain its purpose, along the lines of the following statement:

> The response matrix on the following page is offered for convenience in verifying that this proposal is completely responsive to all requests and specifications in the RFP, either implied or explicit, and to evaluating the proposal accordingly.

Transmittal Letter

Proposals are self-explanatory, but it is customary and a matter of protocol to enclose a transmittal letter with a proposal, as in Figure 6.6.

Capability Brochure

There is a need in many fields of contracting for a marketing document referred to variously as a capability brochure or capability statement, and may even be referred to as a qualifications brochure or statement, for that is what it is, in the final analysis. In fact, while it is a requirement stipulated by clients in many cases, it is more than that. It is an excellent tool to have in general for marketing contracting services of most kinds. It can be a useful marketing tool that you furnish

FIGURE 6.6 A Model Transmittal Letter

Dear Major Moriarity:

Enclosed herewith is our proposal for the reverse engineering of the MystoCoder Communications Set to produce a complete set of "as built" engineering drawings. In this proposal you will find detailed plans for security, technical management, contract administration and cost control, all matters that were addressed carefully and stressed heavily in the RFP and in the list of evaluation criteria.

Our proposal also points out our extensive experience in this kind of work, with many similar projects carried out successfully for the U.S. Army on earlier occasions over the past 20 years.

I have been authorized by our corporation to make this offer on behalf of the corporation and to assure you that it is a firm offer for a minimum of 120 days from this date.

Should you or the government's staff require additional information, either spontaneously or in formal presentation, we shall be pleased to comply promptly.

Sincerely,

Martha Siegal
Chief Engineer

voluntarily when a client has not required that you furnish such a statement.

That is especially the case where one contracts for highly customized or nonstandard services. Look at it from the client's viewpoint: He or she wants something done for which there is really no standard. (Even writing a manual or newsletter falls into that class, as each effort is unique.) How is a client to know who is best qualified or most capable to handle the job well? The client may very well wish to limit the request for bids or proposals only to those contractors who appear to him or her to be well qualified and entirely capable of doing the job well. Such a statement may therefore qualify you to compete for a contract that otherwise might be an inaccessible marketing opportunity.

The capability the client seeks may be of more than one possible kind. It may be a technical capability, a capability to provide the staff required, a capability of adequate equipment or other physical resources or even some other kind of capability. (Government contracting officers, for example, may seek to establish a potential contractor's financial resources, where that is a consideration.)

Because there are such variables in the possible definition of "capability" as applied to a given situation, it is difficult to create and maintain a standard statement or definition that covers all cases. Even if you could and would create a sizable capability brochure that covered all possible cases, the client who wants to be assured that you have the proper equipment to get the job done does not wish to wade through page after page of irrelevant information. Thus, while you may and probably will have a general brochure that describes what you do and your general qualifications, you often need to develop and submit a special capability statement to persuade a potential client to invite your bid or proposal (probably the latter).

You "met" Hackney & Williams Software, Inc., and Accurate Litho earlier in this chapter, and saw examples of the proposals they might and probably would have written in pursuit of contracts, responding to the client's request for proposals. Their proposals of course were directed specifically to the clients' specific requests and statements of work. However, the fourth chapter of each one's proposal (Figure 6.4) is primarily a statement of the proposer's capabilities as they related to the specific project and contract for which the proposer is competing. Thus, with relatively little modification, those chapters might have served well as capability statements for those requests, had the client called for such statements.

While it is a good idea to have a general brochure that describes what you do and your qualifications for doing so—education, special training, experience, facilities and other relevant resources—you often need special capability statements to establish your qualifications with a specific client for a specific project.

Here, your computer becomes an invaluable asset: You make up standard paragraphs and sections for the various aspects of your capability, as appropriate—personal, educational, training and experience; equipment; physical facilities; other personnel (e.g., consultants and associates you may call on)—and thereby construct custom capability statements easily and quickly, as the occasion requires. (These are, of course, equally useful as proposal-writing assets.)

As an independent contractor, your objective in all of this is to win contracts. But what kinds of contracts? Which are most suitable for you? Which are least suitable? We'll address those and other contract-related questions in the next chapter.

7

Contracts and Contract Negotiations

A working knowledge of contracts, contract law and contract negotiations is as important to the independent contractor as it is to the working lawyer.

The Meaning of the Term *Contract*

Contracting is an extremely broad and general term, as already noted. In the strictest legal sense, just about every business transaction involves a contract. That is because a contract is any agreement that satisfies five general conditions or requirements:

1. One party makes an offer.
2. The other party accepts the offer.
3. There is a consideration involved in the exchange.
4. The parties are legally competent to enter into agreement.
5. The agreement is to do something that is legal.

Anyone operating a retail establishment is making an offer to sell something. A customer accepts the offer and pays the price. It also works the other way around: the customer makes an offer to buy something and the other party accepts the offer at some agreed-upon price, so the basic conditions still are met. That payment is the consideration, although the consideration does not have to be money. As long as the two parties are of legal age and mentally competent, they may enter into a binding agreement. To be binding, the agreement must be for an action that is within the law. (An agreement to buy stolen goods, for example, would not be a valid contract.) Thus every legitimate purchase normally involves a contract, although the contract may be a tacit or unspoken one.

Simple as this set of conditions is, large volumes have been written about contracts, contract law and directly related issues, although we will not go deeply into the subject here. However, two related points must be made here:

1. The consideration is usually money, but does not have to be money. It can be any other compensation, even an agreement not to do something in exchange for what the other party wants.
2. The contract does not have to be in writing. In fact, it is the agreement, not the written description of the agreement, that is the contract. Verbal contracts are as binding as written ones, despite the remark allegedly made by moviemaker Sam Goldwyn that verbal contracts "are not worth the paper they are written on."

A Few Related Considerations

There is neither need nor room here to get into anything resembling a full-blown discussion of contract law, but to have a practical grasp of contract law, it is necessary to know some of the peripheral matters and principles beyond those already mentioned. Some of these will surface naturally in the course of discussions as we proceed, but first, a few general observations.

There are two kinds of law: statutory and common law. Statutory law is established by legislative acts, such as those of Congress and legislative acts of state and local governments. Common law, sometimes referred to as "judge-made law" or "court-made law," is law based on legal doctrines established by courts and may include also courts' interpretations of statutes. Common law is therefore based on the precedents established by such earlier decisions. When contract disputes arise and the two parties cannot reach agreement, the dispute must be referred to arbitration or, that failing, to court. However, winning a contract dispute in court too often is a Pyrrhic victory, since legal costs can easily become greater than the worth of the contract. To avoid that possibility, many contracts include an arbitration clause, in which the two parties agree to have the matter settled by arbitration, in event of a dispute.

The probability of disputes is greatly elevated when a contract leaves things unstated. The more specific and detailed a contract is about who is to do what, when it is to be done and how much is to be done or delivered, the smaller the probability of a serious dispute. Thus is demonstrated the advisability of a detailed proposal made part of the contract by reference as the statement of what is to be done.

The probability of disputes is elevated even more when reliance is placed on a verbal contract. The problem with verbal contracts is not that they are not binding, but that the parties to a verbal contract must depend on their memories of what was agreed to. In the event of a later dispute, each party's memory is likely to be rather selective and protective, and an arbitrator or court may have great difficulty determining the original intent and agreement. The arbitrator or court will then attempt to determine what reasonably may be inferred as the agreement, and what would have been proper and binding. If, for example, one party was misled through no fault of his or her own as to the intent of the contract—e.g., the other party took unfair advantage of the second party in some way—the contract may be deemed invalid. This particularly would be a possibility when the contract was a verbal one.

It is, of course, not absolutely necessary that every contractual agreement and undertaking be recorded on paper. Most small transactions are not so recorded. It would be impractical to write up an agreement for every small contract, such as having someone mow your lawn or service your automobile. Thus, we do rely on verbal contracts, and even those implied rather than specified contracts.

There are hazards here, even in routine contract work where there is little or no documentation before, during or after the work. When I contracted with a local service station to replace a defective alternator, I didn't think to inquire about a guarantee. I assumed it would be guaranteed for at least 90 days. When it broke down in a few weeks, the contractor, the local service station operator, that is, refused to replace it, except for the full price of another replacement alternator. Of course, he lost my future patronage, but I lost the price of yet another alternator because I did not get a written guarantee and decided I could not spend the time to complain to the authorities to try to pressure this contractor into doing the right thing.

This does not mean that in your business as an independent contractor you must have a contract for every job you do for a client. In fact, there are situations in which you are better off with a handshake agreement. A client requesting a quote on a small job of a few hours' duration and agreeing to your figure may very well become apprehensive if you present some kind of written agreement for the client's signature. The matter of good judgment is very much a requirement here. But before we get too deeply into that aspect of contracting, let's talk about contract types. In fact, we also will be talking about different types of contract services or different modes in which such services are provided.

General Classes and Types of Contracts

Contracts can be classified in several ways. Depending on the type of work for which you contract and the types of clients with whom you do business, you may tend to one type of contract or another. In one sense, there are three broad classes of contracts for services:

1. Fixed price for the job
2. Fixed rates for indefinite quantities
3. Cost reimbursement

Each of these is a general class of contracts, and there may be variants within each class and hybrids among the classes.

Fixed Price for the Job

This is the simplest type of contract. The job and price are specified in their totality. It is useful and appropriate when the job or product can be qualitatively and quantitatively specified, so that it can be demonstrated clearly when the contract has been fulfilled.

Fixed Rates for Indefinite Quantities

Indefinite quantity contracts are common in contracting businesses. Their need is apparent when a client does not know exactly how much he or she will need of something, whether goods or services, but wants to be assured of having the goods or services available for some period of time at some agreed-upon rates. Thus the client may ask you to quote your services by providing a rate schedule and offer to contract with you for some period of time—a year, perhaps—at those rates. (Such contracts are sometimes identified by contractors as "laundry list" contracts, referring to a list of many items of service.)

The exchange here is that you guarantee your availability for the client's needs and the rates you and the client have agreed to, while the client guarantees that you will be called on for all services or products covered by the agreement.

Many variations are possible. The contract may or may not guarantee you some minimum amount of work, and that guarantee may be in dollars or in terms of whatever the rate units are—e.g., hours or days. In some cases, the client wishes to express maximum dollars or maximum units of work in such contracts.

The items enumerated in such contracts may be general—e.g., writing services generally—or specific, listing rates for various writing tasks, such as releases, advertising copy or newsletters. It may also list functions, such as writing, editing, proofreading and makeup. The rates may be by hours or days, or they may be by other work units, such as pages or column inches.

Contracts of this general type are referred to by various names to denote somewhat different applications. Among the names used are T&M, for time and material contracts; BOA, for basic ordering agreements; and TO, for task order contracts. The latter two are virtually the same kind of contracting vehicle. They are also referred to as annual supply or indefinite quantity contracts, but all are variations of the same idea: fixed rates with guarantees to each party. The contractor guarantees the rates and the availability of the services; the client guarantees to use the contractor's services and may guarantee some minimum amount of work.

Under such a contract, the client may call on you to undertake an assignment at any time, reserving the right to terminate the assignment or declare it completed when he or she wishes. However, it is more commonly the practice to issue a task description and ask for an estimate of effort required based on the contracted rates. The client may accept your estimate or disagree with it, in which case there is a need to negotiate the task. When agreement is reached, the client provides a go-ahead in the form of a letter or purchase order.

There are possible problems for you here as the contractor in such an agreement. Since you cannot be sure of when the client will request your services, you cannot be sure that the client suddenly will not ask you to undertake a sizable job at a time when you are already heavily committed to another project. Unless you are confident that you can cope with such situations, you will need an escape clause of some kind. That is also in the client's interest, so that the client has an alternative if you have difficulty handling the assignment. The way to overcome this is to provide that you have a guaranteed first right to the assignment of the task, but the client may turn to another source if you are unable to handle the task. Similarly, if you and the client are unable to reach agreement on the effort required and cost of a given task, you may reject the task and the client may place the work elsewhere.

Cost Reimbursement

The famous cost-plus contracts of the cold war at its height are the best known of the cost-reimbursement contracts, although not the only one. Actually, they were (and are, in those relatively rare cases where they are still used) cost plus fixed fee. The concept was that the contractor was all but insured against loss, and in return was to get a guaranteed small profit as a fee for services. The contract stipulated the direct labor rates, overhead, G&A (general and administrative costs, where the contractor had an established G&A) and fee. If there were other costs, such as those for materials, the contractor was required to bill these at their costs (plus G&A). The amount of labor and other costs might vary, but the fee was fixed for the contract, so that it did not vary, regardless of whether the total size of the contract proved to be greater or less than the original estimates.

Even that type of contract has its variants, with some providing a fixed and award fee. The fixed fee is quite small and the award fee can vary, according to the client's appraisal of how well the contractor has performed. The T&M (time and material) contract may appear to be somewhat similar to the cost-reimbursement contract, but in the T&M rates are "loaded," and so include the burden costs and the profit.

In general, cost-reimbursement contracts are relatively rare today, and the other two types are far more common. For many types of contracting, T&M or BOA types of contracts are the most practical and most common, although many clients prefer the fixed-price contract.

Pros and Cons

Each type of contract has advantages and disadvantages, especially when it is misused—used, that is, where it really is not appropriate for the circumstances. The reason for so many types of contracts is that there are many different kinds of situations, each calling for a contractual agreement suitable to the conditions of the situation.

Fixed-Price Projects

Many clients prefer to contract on the fixed-price basis because their obligation is firmly defined up front, and the budget is clearly established in advance. It is often to your advantage too, to contract on this basis, but only when the circumstances warrant this kind of contract.

In some projects, it is possible to specify in the contract exactly what is to be done, but that is not always the case. Unfortunately, sometimes a client is not highly specific about what he or she wants done, perhaps because he or she does not fully understand his or her problem and needs, but still wants you to provide a firm, fixed-price estimate. These kinds of clients have their counterparts in contractors who are eternal optimists, willing to gamble on a fixed price and confident that they can worry their way through any project and problems that they encounter. This can lead to problems you will wish you never had encountered, of course, if you are optimistic enough to agree to a fixed price when the details of your obligation are less than completely clear. Some clients may be understanding when you explain that you did not anticipate all the difficulties, but you cannot count on it. Having signed a fixed-price contract, you are probably stuck with it. If you estimate the cost at a high enough figure to cover all possible contingencies, you are likely to price yourself out of the job. (That is especially the case where you must bid competitively.) Of course, if you estimate below that figure, you risk losing money on the job.

There are at least two ways to undertake such a project and cope with this problem:

1. Prepare your own set of specifications, if that is possible, and offer to contract on the basis of your specifications. Of course, to do that, you must be reasonably sure that your specifications are correct for the job.
2. If it is not practical to write up your own set of specifications, you might offer to undertake the job on the basis of an hourly or daily fee or rate, in the philosophy of one of the other types of contract. That is, explain the problem to the client and show your client that it is in his or her own best interests to contract on that basis, if the job is to be done right. There are contractors who will bid the job low and then cut corners to get it done for the price. That is not in the client's interests, of course. I have usually found it possible to persuade clients to see the truth of this.

The remaining alternatives are to decline the opportunity to submit your bid or proposal or to submit a price that is high enough to minimize the hazard. That will probably be tantamount to declining to bid at all, since it will probably price you out of competition, so that you will have wasted your time, after all.

Fixed Rate, Indefinite Quantity Projects

The typical fixed rate, indefinite quantity contract, typically the T&M, could be for a single project. The duration might be indefinite, if for a given project, or for a specified period, if the services are continuous. This is a suitable contract vehicle when it is not possible to predict everything that might be needed. A company needing help to produce the manuals or training program required to support their project cannot always predict exactly how long you will be needed, whether the job will require you full time for a given period or whether you will have to work your own efforts into their schedules. Quite often, the best estimates prove to be grossly in error in such circumstances. It's wise to allow for an irregular schedule in such circumstances, but also consider the possibility of overtime. Such jobs often get into schedule difficulties and then require you to work later hours and even weekends to get the job done. Your rate schedule should include overtime rates, even if the client assures you that no overtime will be required.

The BOA or TO contract is usually written for intermittent or occasional support services, each authorized by purchase order as a small, fixed-price job in itself, as explained earlier, priced on the basis of the rates established by the contract. (The estimate called for by the client for each task is for an estimate of hours, functions and materials, if any.)

The problem with both these kinds of contracts, especially the latter ones, is the unpredictability of the work. A one-person independent contractor may have to be an artful dodger to handle the work in a series of typical feast-or-famine situations. On the other hand, they can be profitable assignments, and in cases where the client is a busy organization, the work can be relatively steady. If the contract is for a long term—more than one year— there is the possibility that your rates may be made obsolete by rising prices. In that case, you should work an annual escalation factor into the contract. That is a common practice for such situations and kinds of contracts.

Cost Reimbursement Projects

The cost reimbursement contract is less and less frequently offered today, since it was a contract form favored by government agencies and their prime contractors who often had to offer such contracts to their subcontractors. It was and is a contracting agreement in situations where money is far less important than other considerations. That made it a necessity where the contractor was to do the most basic kind of research and development. In that situation, estimating time and cost to achieve some desired result was exceedingly difficult, and no contractor was willing to gamble on a fixed-price contract of any kind. Doing the documentation/publications support for NASA's Goddard Space Flight Center, for example, was a three-year cost reimbursement type of contract. The overhead rate had a guaranteed ceiling, labor rates for each class of editorial labor or function was set with annual escalation rates and the fee included both a fixed fee and an award fee based on performance. NASA's work was heavily characterized by pioneering research and development, including many chores for the military agencies, to which the documentation was necessarily linked. The lessening of the pressures to advance technology and applications for military usage was greatly lessened with the end of the cold war. That lessening of pressures has greatly reduced the perceived need for such contracts. Still, depending on the kind of work you do and the kinds of clients you pursue, you may encounter occasional opportunity or need for such contracting formats. At the least, you need to be aware that such contract forms exist.

Subcontracts

Opportunities for independent contractors in many cases are especially abundant in the subcontracting field, supporting the prime contractors, as many earlier references to subcontracting have pointed out. From your viewpoint, it normally makes little difference whether your client is the end user of your services or a prime contractor. The agreement or contract you sign with a client is the same. However, it is sometimes necessary to understand that the type of contract chosen by your client and the terms your client insist on including are influenced by a prime contract your client holds. Sometimes a prime contract will require that all subcontracts under the prime contract be negotiated and awarded competitively, for example. If a prime contractor has a T&M or BOA kind of prime contract, he or she may want to call on you for help when an occasional overload of work occurs.

Understanding these kinds of circumstances is a help in assessing market opportunities and in negotiating contracts.

Contract Forms

Contract law is a large element of the body of legal codes and provisions, as it should be, considering how much of everyday activity is conducted under contracts that are implied, inferred, verbal, written, simple and complex. All follow a simple set of principles, presented at the beginning of this chapter. That simple set of principles finds its expression in a wide variety of contracts, from the very simple and informal to the massive and laboriously worked out under many hours of negotiations.

Basic Form and Format

The basic form of a contract is presented in Figure 7.1. As should all contracts, it must follow the five principles described at the beginning of this chapter (six, in some cases, where the law requires that the contract be in a written form). It should identify the two parties clearly, giving names and addresses, and should describe what each party is to do, giving both qualitative and quantitative measures, with due dates and payment specified. It's a good idea to print or type each party's name as well as have all signatures, since signatures may be hard to read.

What is to be done—the specification of services to be performed or products to be delivered—may be specified in the body of the contract or it may be incorporated as an attachment (often referred to as an exhibit) or by reference.

Incorporation by reference is a labor-saving action. When you have bid or made a proposal to a client, your bid or proposal—the specification of what is to be done or delivered and the agreed-upon price—may be made part of the contract by reference. That saves the parties a great deal of labor when the specification is a long and complex one, and minimizes the probability of typographic and other errors that so often take place in translation and copying.

The form shown here is the most basic and informal contract model and may even be in the form of a letter. As such, it may be adequate for minor projects, but it falls short where larger projects are involved, and

FIGURE 7.1 The Basic Agreement/Contract Form and Format

Agreement

This agreement is entered into this _____ day of _____, 199 _____
between_____
of_____
and _____
of_____ .

The parties agree as follows:

This is the entire agreement between the parties and may be amended
or altered only in writing, signed by both parties. This agreement is governed by the laws of the State of _____ .

By _____ By _____
   name   name

_____ _____
   signature   signature

---

few clients and even fewer legal representatives would accept this as a
suitable statement of the agreement between the parties.

For any sizable project, the lawyers might insist on having the
signatures of the two contracting parties witnessed, to forestall any
possible later denial of having signed the agreement, but it is normally
not an absolute requirement.

The notary seal is in the same category: It is not an absolute
requirement in most cases, but many parties to contracts, and especially
their lawyers, feel more secure if the contract is witnessed and the
signatures sworn to by a notary.

The Purchase Order

For small projects, many organizations use a purchase order form, such as that in Figure 7.2. It is a simplified contract form, usually suitable for small projects that have no complex requirements. The client may simply type a brief description of what is to be done or delivered, with a due date. You sign and return one or more copies, and a contract exists between you and your client. Where the space is provided for your signature is listed here as "signed," but it means "accepted," and that word is, in fact, used in some purchase orders.

Contract Issues

A great many issues and questions may arise in connection with contracts. Here are a few, beginning with and elaborating on some of those already mentioned:

- What, precisely, is to be done? To what standard of quality and in what measure will it be done, and when will it be complete or delivered?
- What is the consideration? When and how is it to be paid?
- What state's laws govern this contract and its interpretation?
- What measures, other than lawsuit, shall be instituted to resolve disputes if they cannot be resolved between the contracting parties?
- Is the contract to be declared null and void if one clause is violated or can the contract survive such an incident?
- Who will be made privy to information the other party deems to be confidential?
- If you are subcontracting and working directly with the prime contractor's client, should you be enjoined from doing future business directly with the prime contractor's client?
- Are you in fact (technically, that is) an independent contractor or a temporary employee of your client?
- Can you assign your contract or receivables to another party?
- Do you retain any rights of ownership in a product you develop for a client?
- Can either of you terminate the contract before the scheduled date?

FIGURE 7.2 A Typical Purchase Order Form

Excelsior Electronics, Inc. **1724 Industrial Avenue** **Voltage, Ohio 45333** **555-675-6657**	**Purchase Order** **Date:**
Issued to: _____ _____ _____ **Telephone Number:**	**Authorized by:** **Ship via:** **Ship to attn:** **by (date):**

Description	Price	

Purchase order number must appear on all invoices and correspondence. Please sign and return second and third copies.

Signed: **Date:**

Contract Negotiation

The best time to solve a problem is before it arises—i.e., via preventative measures. The objective for negotiation is to reach a clear and precise agreement, on the one hand, but also to try to anticipate all possible problems and make provisions to either prevent them or to have an agreed-upon means ready at hand to address and solve the problems. To negotiate intelligently, you must be prepared with an understanding of the issues. (In fact, the more obvious it is to the other party that you do understand the issues, the more tractable the other party is likely to be.)

Why Must You Understand the Issues?

As a lay person, you cannot rely on common sense—everyday logic—as a means for anticipating typical contract or project problems and providing for them in negotiating a contract. Nor can you be expected to understand the significance of such clauses if you use preprinted standard contract forms, such as you can buy in any good stationery or business supplies emporium. It is a good idea to read the following brief explanations of these typical contract clauses, since the explanations will help you understand what those clauses mean and why the other party may insist on including some of them in any contracts between the two of you. Certainly, if you do not have at least a general understanding of contract law and these common clauses, you will be at a serious disadvantage in contract negotiation even if they are of the most informal kind.

Specifications

It is not easy to establish a standard of quality in most contracting work. A generally accepted standard in most industries is often referred to in contracts as "good commercial practice." That means conformance with what those experienced in the industry or with the services and products involved normally deem to be acceptable. In some cases, such as in some construction trades—for example, electrical and plumbing work—there are building codes, enforced by inspectors who must approve or disapprove the work. In many cases, however, there are no set standards or at least none that can be measured objectively. If you construct an oscilloscope or write a technical manual for a military organization, the contract may invoke existing detailed military standards and specifications by which the conformance of the work and product can be measured. But even the military does not always require that things be "built to spec" when their budgets are tight and they are asking for lowest prices. Then, they must make objective evaluations. There are at least two methods for dealing with the problem of quality standards when exact standards or measurable specifications are unavailable:

1. One method for minimizing the probability of serious dispute about standards of performance or product is to provide for periodic product or service reviews by the client to ensure that if there is to be a dispute, the problem can be remedied early on by negotiating any differences immediately. That is a common method of handling the problem in custom work. Find the logical milestones in the project—e.g., completion of a first draft in a writing project, or construction of a laboratory model in a hardware development—and provide for client inspections and reviews, with adjustments to be made immediately, in the event of objections to what you have done so far.
2. Another method, probably less satisfactory, is relying on "good commercial practice." This may mean, in the event of a dispute, finding someone who is satisfactory to both parties as an expert, and agreeing to be guided by his or her judgment and decisions.

Consideration

The consideration is usually some fixed sum of money, and that is easily measurable. How it is to be paid can lead to disputes, however. If the project is to be a long-term one, you probably will want to be paid periodically, as progress payments, and it is reasonable to insist on this. But there are other concerns: What happens if your client is delinquent in making payments? You may choose to suspend work until the client catches up with payments, but the client may deem that to be breach of contract. Thus, the right approach is to agree on a formula for coping with the situation and include it in the contract.

Governing Law

If you and your client reside in the same state, presumably that state's laws will govern all disputes and settlements. But suppose your client resides in another state? You will have to reach agreement on which state's laws govern. That should be the subject of a clause in the contract.

Settling Disputes

When a dispute arises on which you and your client become deadlocked and thus cannot settle between the two of you, what will you do? Go to court? There is a better way: You can agree in a contract clause to submit such disputes to arbitration, and you would usually settle on a branch of the American Arbitration Association in the same state in which you have agreed is that of governing law.

Violations versus Contract Invalidation

The violation of a single clause in a contract can be construed to be cause for invalidating the entire contract. That is not always desirable, however, especially when the violation is a technical one and not of great importance. To erect a safeguard against such disastrous outcomes, you can insert a clause that a single violation will not invalidate the remainder of the contract.

Confidentiality

The confidentiality clause is one of the most common clauses included in contracts. Contracting often means establishing a close working relationship with your client, so that each of you becomes privy to information belonging to the other. This may include technical data, operational methods, marketing information, financial data or many other data regarded as trade secrets and other classes of proprietary and confidential information. It is quite common to include contract clauses in which each party pledges to safeguard the other's proprietary information.

Noncompete Clauses

Many contractors accept subcontracts that require doing work on the premises of the prime contractor's client. The prime contractor has two concerns in this regard:

1. That you, the subcontractor, may attempt to do business directly with the client in the future, stealing the prime contractor's client.
2. That you may accept a permanent position with the client, again at the expense of the prime contractor's business assets.

To protect against these contingencies, the prime contractor is likely to ask you to agree to a noncompete clause in the contract. In this clause, you agree that you will not go to work directly for the client, either as an independent contractor or as an employee.

Such clauses are quite commonplace and usually are effective for one or two years from the date of the contract. Occasionally, a prime contractor will try to persuade you to agree to a much longer period, but two years is more than adequate and ought to be the maximum to which you will agree.

Independent Contractor Clause

Because of the resistance of the IRS to individuals claiming independent contractor status, many independent contractors insert a clause in their contracts affirming that they are independent contractors and not employees and are so recognized by their clients, the other parties to the contracts. The clause is far from a complete guarantee that the IRS will accept the statement, but it certainly does no harm to have such a clause in your contracts.

Assignments

Ordinarily, you can assign—convey to someone else—anything that is your property to begin with. It can be almost anything, such as a contract, part of a contract, money due you or other assets. Clients are not always fond of finding that they are working with another contractor than the one with whom they originally signed, and so often include a clause that forbids any assignment, with the probable exception of receivables. Ordinarily, they will not object strongly to your assignment of your receivables to a bank or other factor. It is in your interest to be able to do so, and you should insist on keeping that right.

Rights in Product

The matter of who owns a product you develop—e.g., a computer program—may arise. In most cases, the client claims and enjoys the exclusive right of ownership in a product you create as the client's agent—his or her contractor. However, you may have an exceptional case. There are at least two such possible circumstances:

1. You may propose to use, in carrying out the project, a product or part of a product you already own. In such case, you need to make that clear in the contract so that you do not compromise your ownership.
2. You may have trouble getting the price you want, and be able to use ownership of the resulting product as a bargaining chip. That is, you will work at a lower figure in return for retaining the right for further use or sale of the product you create under contract.

You would thus grant the client a nonexclusive right to the unlimited use of the product, but perhaps limit that use to his or her own business, forbidding resale of the product. The reverse may also be the case, with you retaining a nonexclusive license to use the product and possibly to sell it.

Termination

Every contract has a normal termination upon completion of what is specified in the contract. In some situations, however, termination is a bit cloudy, where you or the client are unsure about the future, with regard to the contract. Thus you may wish to agree that one or both of you may terminate the contract under some specified conditions.

Armed with this kind of information, you will be much better prepared to interpret, write or negotiate contracts.

Among the most common dilemmas confronting beginners in independent contracting of all kinds are those of pricing their work and handling problem situations from the need to set and quote prices, including price negotiations.

8

Setting Your Rates and
Negotiating Them with Clients

*The price you place on your services should be set arbitrarily. Your
decision must be guided by your needs and conditions if you are to
survive, much less prosper.*

A Troublesome Matter

Deciding how to price services is a troublesome matter for many, perhaps even most, individuals selling services of some kind. It is one of the
earliest matters of concern and bewilderment to a great many of those
newly launched into an independent contracting venture, but it is not
confined to only those new to business; it also troubles many of those
well established and reasonably experienced in their contracting ventures. They continue to wonder whether they are pricing their services
wisely or competitively, whether they ought to raise rates to be more
profitable or lower them to be more competitive. They debate these
issues in whatever forums they can find that lend themselves to such
discussions, such as electronic (computer) bulletin boards and associations to which they belong. Less experienced service entrepreneurs,
however, do not debate as much as they seek advice from others in
similar enterprises.

Much of the advice given to those who seek guidance in this area of
business operation and policy does not make a great deal of sense. A
common guideline offered by many is to base one's prices on what one
wishes to earn or earned previously as someone's employee. Is that a
sensible basis on which to base your rates? Is it likely that the customer

or the market is at all concerned with or would be influenced by your wishes or salary history? Of course not! If that were useful advice—if we were free to set our rates as high as we wished—we would all charge the most outrageous fees, if we could find clients willing to pay those fees. Of course, we could not.

The Basic Ways To Price What You Do

There are only two practical bases on which to price your work as an independent contractor:

1. Set a total price for the job.
2. Set a rate for each unit of your time (hour or day).

Which method you use depends on several factors: your personal preference, the nature of the work you do and the individual circumstances of the given situation. In many professions, it is customary to charge by the hour or portion of an hour. I once got a bill from my lawyer listing every telephone call he made in my behalf, along with the portion of an hour involved. In some other professions, it is customary to charge by the day.

Actually, there is a third way to charge for what you do. That is to set a fixed rate for certain specific services—e.g., to hang a door, write a résumé or conduct a seminar—if your work lends itself to such unit pricing by the standard task.

You are not, of course, compelled to use any of these methods. You can choose among them for yourself, use any combination of them, hybridize them in some way or invent an entirely new way to price what you do.

The Trouble with Setting Rates

Setting rates for your services appears on its face to be a simple enough proposition. Presumably, you set your rates to meet these goals:

- Your rates are competitive.
- You are able to pay your bills, the normal business expenses.
- You are able to pay yourself a reasonable salary.
- Your venture shows a profit.

That seems simple enough. The problem is that it never is that simple. The whole question of pricing is complex, involving a great many considerations and miscellaneous related aspects of your business.

The Established Market

A possible starter point for setting rates is in evaluating the "market"—the established and generally accepted market value—for your product or service. The market is not a fixed price, however, it is a range of prices. It does, however, set normal limits that are commonly accepted by entrepreneurs and clients.

Suppose you find, for example, that most contractors charge from $12 to $16 to install a glass tub or shower enclosure. That then defines the market generally for that service. It means that you ought not to have to go below $12 to sell such an installation, but you can as high as $16 without becoming "noncompetitive" if you subscribe to that definition of the term.

Unfortunately, the most common interpretation of "competitive pricing" is pricing that is as low as or even lower than the prices charged by others for similar goods and services. Certainly, that factor of what your direct competitors charge should be considered, but it is far from being the sole or dominant factor.

Pricing and Marketing

Earlier in this book, I emphasized that marketing is a most important element, perhaps *the* most important element bearing on your success as an independent contractor. Obviously, pricing is also an important element of your business. In fact, pricing is an important aspect of marketing, as well as a function of accounting.

Many business owners use pricing as their main sales strategy by offering or claiming to offer bargain prices to attract customers. This rather obvious way to use pricing to enhance sales is probably the most commonly used sales strategy. Such words as *sale, bargain, discount* and similar terms never lose their appeal. However, price appeal can go the other way too, because many customers are impressed by high prices, assuming them to be a mark of greater quality than that offered by other, competitive services. It may or may not be accurate as an indicator of quality and dependability but, given certain qualifying conditions, higher prices than your competitors charge their clients may or may not cost you clients immediately, but high prices sometimes produce a long-range effect.

I was privileged to witness an interaction recently in which one computer consultant, whom I shall call George, discussed his rates with a number of other computer consultants. George remarked that soon after starting his consulting practice, he had decided that he would never make more than a modest living at the rates he was charging clients. He therefore raised his rates sharply, and listed them for the others.

One of his listeners responded that if he raised his rates that much, he would lose one-half of his clients immediately. George responded that he had, indeed, lost one-half of his clients when he raised his rates. But that, he said, freed him up to replace those clients with clients similar to the remaining one-half, clients who valued what he did enough to pay his higher rates.

The point is that there are clients for bargain rates, and there are clients for high rates. The difference is in how the clients perceive you and what you offer. There are all kinds of people, so inevitably there are all kinds of clients. Some will judge you by your price and will see you as poor, mediocre or excellent, according to your rates or fees. Some will always seek the lowest price, while others will be repelled or even made suspicious by what they think is too low a price. Some can be educated to see that you and your service are worth more than many others charge, while others will resist all sales presentations.

A Matter of Value?

How does one establish a value—hence a price—for any custom service? Bear in mind that a custom service is a unique one, much as it may resemble services offered by others, so comparisons with competitive services and the rates charged for them offer only the most approximate guidelines. Anyone who has called upon different services to repair an automobile, mow a lawn or clean a carpet soon realizes that there are distinct differences of degree and quality of service between and among those who purportedly offer similar services. For this reason alone, many of us deliberately and knowingly engage the higher-priced services. I know that I pay premium prices to have my automobile serviced by the dealer, rather than the local service station, but I find more than one reason to opt for the dealer's service, despite the higher prices. It's true for all kinds of services. Many doctors, lawyers, consultants and specialists of all kinds charge huge fees and are kept busy because they are reputed to be worth their fees.

Value is difficult to define. In this context, the client ultimately defines it, much as we try to shape the client's perception. A piece of art for which I might not pay $10 may be worth several thousands to someone else. The job of marketing is not only to win sales, but also to get sales at the prices you have decided you must get. Thus setting value—shaping the client's perception—is one of the functions of marketing.

Profit Must Be Considered

There is an ancient platitude in the business world to the effect that you will never go broke while taking a profit. Like many other platitudes and clichés—and the business world has its own large supply of both— this clever little aphorism has a germ of truth in it, but can get you into serious trouble if you take it too literally.

For one thing, what is profit? What does that word really mean? It is not as easy to define as it may appear to be. Today, *profit* is a noun that needs an adjective to make its meaning clear. In the company of such modifiers as *gross, net* and *pre-tax* the word *profit* takes on many different meanings. Realizing a gross profit, the difference between what you paid for something and what you sold it to someone else for, doesn't guarantee you a net profit, by any means. We will define those terms later in this book. But there is another side to the entire question. There is the matter of how pricing relates to other major business functions, such as accounting and marketing, for there is an interrelationship that must be considered.

The Fundamental Fiscal Balance

A common mistake most of us make is to assume that the other fellow knows what is obvious to us. I once conducted one of my seminars on preparing proposals for grants and contracts for a nonprofit company in northern New Jersey. When I reached the subject of cost estimating and discussed overhead, I was asked what I thought was a fair and ordinarily acceptable rate of overhead for a nonprofit corporation. I suggested that anywhere from 50 to 100 percent would be in the ballpark, with 50 to 65 percent reasonably competitive. The sudden silence that fell over the room was one of shock. Mouths were agape, eyes wide and aborted questions hung in midair.

My questioner finally recovered enough to ask weakly, "You mean that six percent is too low?"

It was my turn to respond with a stunned silence. Finally, I was able to get it out: "Six percent! How could you possibly operate even a nonprofit at six percent? Fringe benefits alone cost more than that. Where does the money come from to pay for rent, heat, light, taxes and insurance?" (I nearly added, "And my fee?" but caught myself in time.)

After we had all recovered, further discussion brought forth the information that this organization usually charged about 25 percent overhead, but had been all but required at times to charge 12 percent, and even 6 percent, to win the grants it pursued.

Of course, I wanted to know how it managed to pay its indirect expenses. After some probing, I learned that it did so by getting certain state subsidy payments to make up what would otherwise have been losses or financial shortfalls.

I assume that you do not have that kind of an "angel" to bankroll your deficits, and that your business venture must pay its own way. We keep books to tell us what the score is at all times because it is not always easy to tell whether your business is paying its way. Businesses get into trouble constantly because their owners do not keep close track of income and outgo, and one day they suddenly discover that they are far too many dollars in debt and may not be able to crawl out of the hole at all. That is why you must know just how things stand on your books at all times.

Understanding the Principles

A problem is that most nonaccountants do not understand the books: we know nothing of accounting. For that reason, I was forced to devise a simple system for analyzing a fiscal problem when I was called on to do so. (The president of the company that employed me as an executive at the time did not understand the comptroller's analysis of a certain project and assigned me the task of studying it and offering my recommendation on the advisability of pursuing the project.)

I did something that would no doubt make any self-respecting accountant shudder violently: I took a sheet of paper and drew a line in the middle, from top to bottom, creating two columns. I labeled the left-hand column *outgo* and the right-hand column *income*. In the left-hand column I listed all the items we would have to pay out, with their amounts. In the right-hand column I listed all the receivables we could count on receiving, with their amounts. I then totaled each column.

The fundamental balance—or lack of balance—between outgo and income is easy enough to understand. If you cannot see enough dollars coming in to pay all those dollars going out, you do not have a viable proposition. You may be able to make it viable by reducing the number of dollars going out, if that does not reduce the number of dollars coming in. You also may be able to find a means for increasing the number of dollars coming in. Or you may be able to do both and achieve an imbalance that is the kind you are looking for: more dollars in than out.

A professional accountant would consider this method an oversimplification of a complex subject, I am sure, and would be right. Accounting is not quite that simple, as you will see in a moment.

The typical accounting system lists dollars in and dollars out under numerous titles, classes and categories. We lay people cannot always see clearly which dollars are those we are getting and which are those we have to hand over to someone else—that is, we can't always tell income from outgo.

To make matters worse, not every dollar spent is outgo or cost. If we spend dollars for inventory, those dollars are neither in nor out; they are merely dollars we have converted to another form; instead of $1,000 in the bank, they are now $1,000 worth of something we shall use or sell eventually. An accountant might say that we have exchanged one asset (dollars) for an asset of equal value (inventory) so nothing has changed, as far as our books and financial condition are concerned.

If you spend $10,000 for a new piece of equipment, you pay the money out, but it is an investment. Again, you have exchanged your dollars for another asset. You will depreciate that asset each year, so it will represent a cost equal to that year's depreciation. However, you may "expense" the item (treat the whole amount as a one-time expense) if it is not too large (the current year's tax code will tell you how large an amount you can expense this way).

For this reason, I recommend a simple accounting system, such as that illustrated in Figure 2.1. A simple system such as this is adequate for an independent contracting business; you can maintain the system yourself, and it helps you understand on a continuing basis just what is happening and how your income and outgo compare with each other.

Overhead: The Enemy of Profit

Overhead, the cost of doing business, is pure cost, total outgo. It can be a modest cost or it can be an excessively inflated cost. When it goes out of control, as it often does, it can do enormous damage to a business.

Overhead is a term used to identify all those business costs that go on whether you are doing business—i.e., making sales—or not. That would include rent or mortgage costs, heat, light and other costs, both fixed and variable.

There are two ways to control and, especially, minimize overhead. The first and best way is prevention, the avoidance of overhead costs. Think carefully about how big a Yellow Pages advertisement you can afford, or whether you should have such an advertisement at all. Remember that such an advertisement becomes a fixed cost for that year. Be especially careful about undertaking costs before you even start taking in money. There is a good bit to be said for starting modestly. When I launched a shower-door installation service many years ago, my first and only initial investment was $10 for a down payment on a truck. I made do with whatever hand tools I owned, and bought better tools and whatever else I wanted or needed out of my earnings, as I went along. (I have done that since, in running an editorial-services business from my home, buying ever better and more sophisticated equipment only as earnings from the business made the purchases possible.)

A Common Beginner's Mistake

One very common mistake many beginners in such enterprises as independent contracting make is to confuse salary or personal draw with profit. The salary you pay yourself is part of cost. You are and should regard yourself as the employee of your business venture. Profit is what is left after all costs, including your own salary, are paid. Profit is what you use to help your business grow. Use it to buy the equipment, furniture, fixtures and supplies you need. Use it for necessary maintenance. Use it for expansion. Use it to carry you and your venture through slack periods. Profit is a must. Without it, you will not be able to do any of these things, except at great personal sacrifice.

A Methodology for Setting Rates

Here are three guidelines recommended by some for setting hourly rates:

1. Determine the typical annual salary for a staff employee doing the kind of work you do as an independent contractor.
2. Divide that figure by 2,080 (52 weeks × 40 hours/week) to derive the employee's hourly rate.
3. Multiply that by 2.5 to arrive at an hourly billing rate.

Thus an annual salary of $35,000 becomes an hourly billing rate of $43.75 (17.50 × 2.5), at least nominally. (You would probably round that figure off to $40 or $45.)

The difference between $17.50 and $43.75 is to cover the fringe benefits an employee gets and you must provide for yourself, and the guarantee of 40 hours pay per week, which the employee gets and you do not get.

That may seem to you to be a most handsome rate of pay to get, even when you must pay for your own insurance and other fringe benefits, but consider this: You are not guaranteed 40 billable hours per week, and if you work 40 billable hours per week it will be at a cost of probably 10 to 15 unbilled hours. They will be hours you will spend doing administrative work, marketing and other necessary chores that are part of your overhead cost. You will thus be billing probably 25 to 30 hours per week at best, on the average. You may get some long-term assignments where you can put in 40 billable hours every week, but then there will be many days when you have no assignment and thus no billing at all. In fact, that $45 hourly rate may prove to be a very modest one.

Once Again, However, What Is "the Market"?

Of course, you cannot neglect the going market for the work you do. It is necessary to determine what your competitors are charging. There are a number of ways to do this. To compare, check the competition informally by chatting with people in the field and in your local area because rates may be quite different in other areas. You may be able to find a friendly contemporary or join a local association where you can get the information you want.

You may discover that the rate at which you have arrived is not competitive, but is higher than most others charge. Will you reduce your asking price or stick with it? It's up to you, of course. But I recommend that you first do the following to figure your costs:

- Review your business plan. (I hope you did write one.)
- Make realistic calculations to see what those necessary fringe benefits and other overhead—rent, light, heat, insurance, taxes and other costs are likely to be.
- Add them up.
- Figure a reasonable salary for yourself and add that to the previous subtotal.
- Do all this on an annual basis to get a final total of all costs.

Figure that two-thirds of your time will be billable. That will be 2,080 × .66, and that will yield 1,372.8 hours. Round it off to 1,400 hours, a reasonable figure. Multiply that by your tentative hourly rate. Call it $45, and it will equal an annual $63,000. Deducting your salary of $35,000 (if that is, indeed, the salary you have decided to settle for) leaves $28,000 to pay all other costs and, hopefully, leave a bit of profit. If your overhead has been, say $20,000, a reasonable 57 percent overhead rate, you will have a substantial sum, $8,000 as an annual profit. You can use this to reduce your hourly billing rate, if you think it is high for your area, to increase your own salary or to save as net profit.

It is an excellent idea to calculate a tentative hourly billing rate by both methods. Each will then serve as a check on the other and increase your confidence in the rate you will charge.

Other Concerns and Considerations

Since you are an independent contractor and in charge of everything, you can choose to believe that you must earn more money, that you must pay yourself $50,000 a year, for example. In that case, the numbers used earlier will not add up. A gross annual income of $63,000 will not support a $50,000 salary unless you can manage to pay all other expenses, including all taxes, with the remaining $13,000. That would mean a 26-percent overhead rate, at most, which is almost surely a bit too lean. You should keep your overhead as lean as possible, but there are limits to how low a figure is possible.

There is another philosophy, which is time-honored and has been offered by many successful entrepreneurs as sage and sound advice. It is based on the widely accepted premise that it is a feat to manage even a break-even the first or second year, much less become profitable before three or four years. Businesses rarely spring to instant or near-instant success. They usually require several years of careful planning to secure every possible advantage.

The objective of starting a business is not to give you an immediate windfall or boost in your regular earnings, but to build something for the future. This usually calls for sacrifice, as well as hard work, at first. It translates into keeping all overhead expenses, including your personal draw or salary, to the minimum. Successful entrepreneurs rarely take a penny more out of their businesses in the early years than they really need. They work at keeping their personal expenses as small as possible, to that end.

In this case, if you can manage to survive on a personal draw of less than $35,000, you give your business that much more of an edge to succeed. You can put aside a greater profit as a hedge against future slow periods and unexpected needs. If it is necessary, you can afford to offer low rates and more marketing activity to bring in the sales that you need to build a business on firm ground.

Should You Have Variable Labor Rates?

Some contractors have more than one rate or set of rates for their labor. An argument can be made for a single rate, but one can also be made for multiple rates. There are several bases on which various rates can be established. For varying labor rates, one might base prices on the following:

- Type of work
- Length or size of contract
- Type of client
- Special arrangements (e.g., retainer or annual indefinite quantity contract)
- "Wholesale" discounting or subcontracting
- Special conditions

Each of these merits discussion. There is not necessarily a logical basis for the special rates in each case, but your decision need not be based on logic. Some may base an argument on ethical grounds also, but these too are often a matter of individual judgment.

Type of Work Some contractors who offer a wide variety of services believe that they are entitled to charge more for the highly skilled and highly specialized tasks than they do for the less-skilled ones. A contractor who consults, lectures and trains others may feel that each of these activities merits a different hourly or daily rate, for example.

Length and Size of Contract A long-term contract, one that provides full-time billing for many months or even for years, puts you in a favorable position with regard to your overhead. It eliminates idle time during that period and probably cuts marketing costs drastically during that same time. You may find it expedient to offer a special, lower rate to attract clients for that kind of contract. Or, conversely, the client with a long-term contract to offer may insist on negotiating a more favorable labor rate. On the other hand, some contractors will charge a special, premium rate for the truly short-term contracts, especially those where the client is in some panic and needs help on an emergency basis.

Type of Client Some contractors will set a lower rate for the large corporation than for the smaller clients. That is not necessarily arbitrary, either. Those organizations who use contractors frequently often have rate scales of their own, and will award contracts based on those rates only. Too, some contractors will cut their rates to get a foot in the door with clients they especially want to win.

Special Arrangements The client who uses contract services regularly or who expects to have need for contract services for a long period may offer an annual supply, indefinite quantity contract, asking for competitive rates as a major factor in the award decision. That situation is similar to some of those already described and referred to. However, there is also the client who offers a nonrefundable retainer as a key ingredient of such a contract or as insurance that you will be available if and when called upon.

Wholesale Discounting or Subcontracting If you are subcontracting your work, you will, of course, have to discount your normal rates by some amount to permit the prime contractor to earn a profit.

Special Conditions If an assignment can be carried out only under conditions that incur a bit of hardship—e.g., being away from home for some period, working under harsh conditions or braving some danger— you are no doubt entitled to set and expect to earn a premium rate.

Should You Penalize Yourself for Being Efficient?

One underlying problem with hourly rates in general for independent practitioners of any sort is that they often do not represent a fair value to the contractor or the client. If you happen to be very practiced and efficient at what you do, the client is probably getting a bargain. You get a job done in less time and at less cost than your competitors. Conversely, if you are slow at getting the work done, the client is the loser. Are you not entitled to a higher rate because of that?

Look at it this way: If you spend $5,000 for a new piece of equipment that saves you time and makes you more efficient in that way, are you not entitled to recover the cost of that equipment by charging a higher rate to amortize it?

Of course you are. You bought the equipment to help your business, and your business must pay for it with some increase of income and profitability, or you should not have bought the equipment.

The average client, however, usually will not appreciate the justice of that, and you may have a difficult time making him or her perceive that and agree to the higher rate. Most of us, even those of us who try to be as honest as possible, find it difficult to believe that which is contrary to our own direct interests. We are reluctant to persuade ourselves to believe it.

It would be much fairer to both parties if we were to set a fair price on the results we are obligated to produce and charge clients on that basis, rather than the amount of effort necessary to achieve the results. There is a way to do this, and it is by charging a flat price for the job, regardless of how much effort is required and of how efficiently or inefficiently we are doing our work.

Fixed-Price Project Fees

Some people take a simplistic view of how to set a fixed price or project fee. They believe that the way to set a project fee or fixed price for the job is to simply multiply the hourly fee by the number of hours you estimate the job will take. Thus, if you charge $40 per hour and estimate it will take you 120 hours to do the job, your fixed price for the job is $4,800. Or is it?

There are problems with this simple formula for pricing the fixed fee job. First of all, it is the same problem we have just discussed, the fallacy of prices based on time expended, rather than on results. Nothing has changed! But there are other concerns and considerations in using estimated hours as a sole criterion for pricing a fixed-fee project.

Presumably, in such a project you are committing yourself to deliver some end-product or end-result for the price you quote. You are to be paid on delivery, whether you get the job done in less than those 120 hours or require more than that for the job. Here are a few questions to pose and find answers to before you commit yourself to that figure:

- How confident are you that 120 hours is enough to get the job done?
- How critical is the schedule final date?
- How well do you understand the goal or objective—precisely what you are to accomplish or produce? How clearly has the client stated it in writing, not verbally?
- How serious a blow would it be to you if you had to put in some extra hours? What is your risk or "exposure"?

Based on the answers to these and similar questions, you might find it wise to add a "fudge factor" of perhaps 20 percent to your numbers to protect yourself. (In your eagerness to win the job or anxiety about what your competitors are bidding, you may very well estimate the job a bit too optimistically.)

In practice, several factors ought to enter into arriving at a fixed price for a project. The first is the estimate of hours, but here there ought to be two estimates of hours:

1. An estimate of how many hours it will take *you* to do the job.
2. An estimate of how many hours it *ought* to take to do the job.

That second estimate must be based primarily on your own expert knowledge and experience. In the automobile repair business and a few others, there are "flat rate manuals." These guides specify how long it ought to take to replace brakes, rebuild a transmission or do many other more or less standard tasks of automobile service and repair. Many automobile repair shops use these manuals to set their prices. In most contracting fields, however, we do not have such manuals; we must rely on our own experience and judgment.

Your price ought to be based on that second estimate, rather than the first one. It is what you might reasonably expect most of your competitors to charge for the project, plus other anticipated costs, of course.

What Others Say

We learn from our experience, but we do not have to rely on only our own experience. We can learn from others' experiences also. Following are a few remarks from others:

> Be very careful about pricing. The temptation is to compete by price, especially when you are new and have low overhead. If you compete on price, expect to get clients who shop only price. You have them as clients only until they find the guy who's a few cents cheaper.

> It is difficult to reprice upward once you have a reputation for working cheaply.

> Working hourly can be treacherous. The assumption is that you can't lose money that way, but you can lose big time there also. There are some booby traps in hourly fees. With your profit built into the hourly rate, you must be able to bill all the hours you spend on the job. On a $500 job, writing off even a couple of hours can kill your profit.

> I found that a combination of hourly and fixed price works best for me. I charge the client by the hour for design, but a flat fee for production work. It satisfies the clients because they have something to get a handle on, and satisfies me because I can be flexible where I have no control over time required and be very efficient where I can control productivity.

> I lost money and had problems with clients until I learned to write clear, fair and detailed agreements up front. Now I spell out in detail what my services include, and in even more detail what they do not. Now I legitimately charge for extras I used to have to give away because they were not spelled out. It has also reduced my liability because the contract clearly limits what I agree to be responsible for.

Price Negotiations with Clients

Negotiations are not always formal affairs where both parties sit at a table and trade offers, counteroffers and objections back and forth, as in the case of negotiating formal contracts for large projects. Negotiations are a key element in every sales presentation whenever a buyer shows interest in a seller's offer and wants to know more about the offer and, especially, about the price.

Overcoming "Sticker Shock"

This is the point where many independent contractors feel most inept and vulnerable. You are not alone if you fear that the client will undergo "sticker shock" and may terminate negotiations immediately when he or she hears your price. It is a common fear, a fear that often impels a contractor who is well worth $50 an hour to offer his or her services at $25 an hour.

This is more than a fear; it is a legitimate problem. Clients who are not prepared to hear a price may react with sudden shock, as I did when I learned what it would cost me to keep my automobile running for another year. As other clients do, I knew it would be expensive, but I hoped that this time would be different, and I would get away for much less than the usual cost of my semiannual general service.

The service manager prepared me. He explained all the things that had to be done and all the other things that were not musts but were recommended preventative maintenance. When he finished that lengthy recital, my heart had already sunk into my shoes; I knew what was coming. I was prepared to be shocked.

That is standard sales practice, but there is more. Here are three relevant rules or principles:

1. Do not volunteer the price information. Wait for the client to ask for the price.
2. Do not release the price information too quickly.
3. Prepare the client for the price information.

Now, let's discuss *why* these rules or principles minimize client shock and rejection and enable you to get a fair price for your work.

Price negotiation is part of selling, a key part of the sales presentation. Why should you not quote a price early in the presentation? An anecdote may help illustrate the reason for this injunction.

A stranger walked into my office one day and introduced himself briefly. I invited him to sit down. He told me that he was in quest of a consultant to help him write a proposal, and asked me how much I charged. He asked that abruptly, without preamble. I knew immediately that it was most unlikely that I could do anything but waste time with this gentleman, so I responded with equal lack of preamble, "Five hundred a day."

"Gee," he said, "you fellows must have a union. Everybody asks for five hundred."

That confirmed my impression that he was price shopping, and not a serious prospect for me.

If you quote your price prematurely—i.e., without doing any selling—you create the impression that the only difference between you and your competitors is price. That is usually death to the sale, unless you are, indeed, "the cheapest guy in town."

A cynic has been defined as someone who knows the price of everything and the value of nothing. For purposes of selling it helps to assume that your prospect does not know the value of what you are selling and thus has no sensible basis on which to judge whether your price is a reasonable one. In those circumstances, it is probable that you and what you offer to do will be judged on price alone. That is why you must first explain what it is that you *do* for your clients, the *benefits* of what you do and the various circumstances of your services. You must *sell* yourself and your services before you quote a figure.

After you sell your services, it is a good idea to wait for the prospect to ask your price. The client who does not ask is not seriously interested in what you offer—not yet, anyway. You must do more selling, more explanation of what benefit the client will derive from your work. The client's reaction is feedback, telling you what to do next. Inquiry as to price is a sign of probable serious interest. But there are always those anxious clients who ask the price before you finish your presentation. In these cases, it's a good idea to delay answering the question, and explain why. I have usually said something along these lines: "I will be more than happy to estimate the cost as soon as I know enough about your problem and what I must do to solve it for you." That kind of response does not antagonize the prospect and does delay answering the question until you are ready to answer it.

It was always my practice to charge a day rate, without premiums for overtime, because I made my day rate large enough to cover me for whatever number of hours it took to get the job done or whatever day of the week on which it had to be done. (The nature of proposal writing often makes overtime and weekend work not only necessary but inescapable.) I therefore was always at pains to explain this before shocking my client with my substantial day rate, and few clients appeared to be shocked.

But what do you say to the client who says you are too expensive or more expensive than your competitors? My answer has always been along the lines that, "I can't speak for my competitors and what they offer, but let's consider what I offer," after which I review all the advantages of my service and my guarantees. In my own case, I point out that there is no overtime and no premium time chargeable, regard-less of when and how many hours I work, and so I am probably less expensive than my competitors, in the end! (You have to find the benefits in whatever your working arrangement is.) I may point out that saving a few dollars at the cost of less than the best job is a false economy. I may offer to guarantee a ceiling number of days to be charged, while guaranteeing that I will get the job done. It depends on how I size up the client. Above all, I never knock my competitors, directly or indirectly.

There may be circumstances where you can or should negotiate, especially if the job is to be a large one or of great duration. The basis for the negotiation should be an exchange of benefits. For example, I have on occasion offered to negotiate a special rate for my client if he or she would sign a formal year-long agreement to use my services on all their proposals or to guarantee me a minimum number of days' employ-ment throughout the year.

That alternative may very well be completely impractical for the client's situation. As a bargaining ploy, however, it can be quite effective, certainly more effective than simple, stubborn refusal to budge from your position. It is a "yes, but" kind of response, positive rather than negative, so that it avoids any contribution to polarizing positions. As a bargaining ploy, it usually works to end objections to your normal price. If truly driven to hard bargaining, I have sometimes offered to work at a much lower rate, but with the provision that I would charge the usual premium rates for all overtime, weekend and holiday work. That too has had a dampening effect on the other's objections to my daily rate. Pricing a job that way—on an hourly basis with overtime and weekend work priced as premium time—makes the total cost uncertain and unpre-dictable. On the other hand, the high but fixed daily rate and not-to-ex-ceed guarantee I offered placed a ceiling on costs, while offering the possibility of costs being less than that ceiling. It can be a most effective argument, if yours is ever a situation paralleling that of my own as described here.

9

A Few Common Problems and Their Solutions

As an independent business owner, you must reorient your thinking from that of the employee. It is more than a matter of psychology or attitude, however. It is a matter of problem-solving, often creative problem-solving.

Beware of Murphy's Law

There is good news about becoming an independent business owner. It offers many blessings and benefits. But there is bad news too. It offers many difficulties and problems. Most of the problems are routine, to be expected; others are special and unpredictable, as Murphy explained with his famous law. What is predictable is that problems will arise, anticipated and unanticipated, but especially the latter.

Change comes about in two dimensions:

1. The general, evolutionary change that is inevitable in a dynamic society
2. The change in your own, individual situation that is also inevitable

Adapting to Change Is Itself Often a Problem

Everything changes in the world about us. Many people have difficulty adapting to new conditions. The desktop computer alone, for example, has been responsible for enormous change. Because microcomputer technology is so new and requires learning, some older people resist becoming familiar with and learning to use them. I have known more than one professional writer to insist that they must write their drafts longhand or via typewriter, and then type it into their computers. When the microcomputer arrived on the scene, accountants who resisted learning to use them soon found their opportunities limited.

But perhaps even more significant is the need to adapt to changes in our individual situations. Working in the business world and rising through the ranks to executive positions, I often had occasion to promote others to new, higher positions. I soon observed that many individuals had problems in adapting to their new positions. Editors who became managers, for example, often seemed to have trouble learning the difference between making managerial reviews of manuscripts and editing them.

Old habits resist being unlearned, for most of us. We are, consciously or unconsciously, reluctant to give up that with which we are familiar and comfortable. The need to learn or adapt to that which is new and different is a condition that underlies many problems you will encounter in operating your own independent business. But what is new and different is not necessarily a new and different problem. It may and often is simply a new and different orientation you must adopt as a result of changes in your own situation. The problems may or may not be different problems than those you encountered as an employee, but you must now react to them as a consultant, and that may call for different reactions.

Changing Your Mindset

The resistance to abandoning obsolete habits and reflexes characterizes a difficulty of many who are new to owning and operating an independent business venture: Many continue to think and be governed by an employee outlook. Operating an independent business, however, calls for a different mindset for a number of reasons. You must forget about keeping an eye on the clock, impatient for quitting time, for example. (I have been surprised to learn how many newly independent entrepreneurs believe that they can build a business in a 40-hour work week.) As an independent, small businessperson, you will soon enough learn that you must be much more dedicated to your everyday work than you ever were as someone's employee.

That isn't all of it, either. As an independent contractor, and especially as an independent business owner, you are going to run into a wide variety of new situations, and what to do about them is entirely up to you. You will have to develop a sense of being in charge and will be compelled to make judgments and other decisions. More important, you will find yourself in a wide variety of roles. You must be prepared for this. Your responsibility is different and much greater, since it is an overall responsibility for everything.

Contractors Are Often Also Consultants

As a contractor and, probably, a specialist of some sort, your clients will think of you as an expert, and perhaps call upon you to exercise independent technical judgment or for advice in connection with some work a client needs. That may come as a formal request for a proposal, or it may come as one or more questions asked of you when you make a sales presentation of any kind, but it comes up constantly. In that situation, you are being asked to be a consultant as well as a contractor. That is yet another role you almost surely will be asked to play at times.

Specialist and Generalist

You may very well be highly specialized in your skills or the kinds of tasks you contract to do for clients. You may even specialize to the degree that you can manage to limit the number of roles you must play as a contractor. You cannot change the fact, however, that you own and operate a business venture and so you are a businessperson. At the minimal degree of diversity, you are unavoidably a manager, comptroller, contract administrator, marketer, salesperson and whatever other roles you play in owning a business, any business, confers on you almost automatically. Even if you make full use of vendor services to handle all your bookkeeping and related chores—e.g., invoicing, reconciling bank statements, pursuing slow payers, preparing tax returns and other such chores—you must be cognizant of and make the decisions relevant to all these activities.

Thus, even the most ordinary business functions are as diverse a set as are the problems associated with any business. Specialist that you are, you also are forced to be a generalist in learning to handle all these matters. And as a contractor, especially a one-person contractor, you may encounter some special problems that are peculiar to that independent-contractor status.

Combating the IRS/1706 Problem

The skepticism with which the IRS views the independent business owner status has become a serious problem for many consultants and other independent contractors, as the IRS employs the famous 20 questions, summarized in Chapter 2, to determine whether you are an independent contractor or a temporary employee.

The problem arises because many consultants and other independent contractors have worked on their clients' premises on long-term contracts (as long as five years or more, in some cases), working the same hours as the clients' employees, even punching time clocks, in some cases, and paid on an hourly basis. Often, they work without clear contractual objectives, but simply do each day whatever the client asks them to do. In many cases, especially with large corporations, the client becomes so comfortable with the contractors he or she hired to help temporarily that the contractors remain on indefinitely. If this is a description of how you are working, it will be difficult to distinguish yourself from an employee, based on IRS standards, other than for the fact that you do not enjoy any fringe benefits.

Your Home Office as a Place of Work

If you have set up an office somewhere in your home, you want to be sure that it qualifies as a legitimate (deductible) business expense based on the following two criteria:

1. Your home office must be a dedicated space, used solely for business purposes. It can be a room, garage, attic or even part of a room, but it must be set off from your personal residential space in some manner, with a room divider or screen, in the latter case, to establish its use exclusively for business. You can do work on the kitchen table, if you wish, but you cannot take a deduction for the occasional business use of the kitchen table.
2. The home office must be essential to your business in some manner, such as meeting the following descriptions:
    *   You meet clients there.
    *   You do your work or create your product there.
    *   You spend most of your working time there.
    *   You keep all your records there.
    *   You can manage or administer your business only there.

Your office does not necessarily have to meet all these conditions, but the more of them they do satisfy, the stronger your position as an independent business. Of these, the first three items are most important, and were among the items specifically cited by the Supreme Court as evidence establishing the legitimacy of a home office as a deductible expense.

Other Important Conditions

Depending on the kind of work you do as a contractor, you may not be able to do most of your work in your home office, spend most of your working time there or even meet your clients there. However, to the extent that you can, you should maintain maximum evidence of business use of that office, such as the following practices:

- Don't accept a client's generous offer to use his or her facilities. If your work must be done on the client's premises, don't do your paperwork there. Have the facilities for and do your paperwork— billing, record-keeping, filing, etc.—in your own office. Have your own copier and fax machine or use one in some public facility, being sure to keep dated receipts.
- Don't adhere slavishly to the client's hours. Do everything you can to set your own working hours at variance with those of the client.
- Don't use the client's tools and equipment, except as that use is inescapably part of the work you have contracted for. Bring your own tools and equipment to the job or take the job to your own facility.
- Have a written agreement with the client and be sure that it specifies the mission: the product or result you have contracted to produce.
- Where possible, work under fixed price for the job agreements, rather than on an hourly basis. If the work is long-term with new tasks being ordered frequently, document each task as a separate task separately priced.
- Avoid being in a position where you appear to be taking orders from the client or the client's staff.
- Spend some part of your time marketing, seeking new clients, and document that.
- Try to have more than one client at a time and find a way to divide your time among two or more clients at a time.

The Importance of Perception

In marketing, what prospective clients perceive you to be or what they think is truth is much more important than what you think you are or what is truth: The prospect will always be guided by his or her own perception. The other party's perception (i.e., that of the IRS) is equally important here. What the IRS sees of you is your tax returns, and it is by these that they are likely to measure and judge you. Therefore, try to meet as many of the above principles and practices as you can. Keep an extensive chart of accounts in your accounting system, and reflect directly as many of these as possible in your tax returns. Show multiple clients there and take deductions for all and various kinds of expenses linked to independent business operations, no matter how large or how small, including such items as tool or equipment costs, postage, printing, faxing, messenger and shipping costs and all others.

The more such items are easily visible in your tax returns, the lower the probability that you may attract a skeptical IRS eye. But also keep detailed backup records of all these things so that if the IRS does challenge at some point, you have the documentary evidence to prove that you are, in fact, an independent businessperson and entitled to file as such.

Time Management

Many of us who are self-employed do not manage our time efficiently. It is probably not as much a matter of mismanagement as it is of nonmanagement: We simply do not think about managing time.

I am one of the guilty or half-guilty. I manage my time excellently on a short-term basis, always devising and improvising efficient little schemes for getting the job immediately at hand done. But my time-management efficiency goes into play only when I am working on a specific project, whereas it ought to be working full time to control and schedule all my work more efficiently.

I happen to be able to overcome this deficiency because I am something of a workaholic. I enjoy my work enough to put in many more hours at it than I would if I were somebody's employee or if I were truly efficient and totally practical at managing my time. I am sure that I am incapable of organizing myself so that I would have and stick to a carefully planned time-management scheme. It is simply contrary to my nature to function that way. It is entirely possible that you, too, would find it simply impossible to be that organized. So let's talk about time management as concerns your entire business activity as well as individual business activities and functions.

Overall Time Management

Managing your time on an overall basis, as so many time-management experts urge, means functioning as an employee, in many respects. It means being in your office, wherever that is, promptly at some specific hour of the morning, awake, alert and ready for work. It means working steadily (perhaps being the superemployee!) for whatever hours you have set as your work day. It means handling all duties efficiently: Handle all papers once only by doing to or with them whatever is necessary the first time you handle them. (Keep a clean in-basket.) It means taking all phone calls as and when they arrive, and thus disposing of them immediately. It means making decisions at once, unless they require further data and study, but a rigorous application of that criterion: Do not rationalize reasons to postpone decisions that do not require delay. It means taking not more than one hour or one half-hour for lunch (whichever you earlier decided was appropriate). It means resisting the appeals from your mate to "take a break" and go shopping or out to lunch somewhere. It means resisting the appeals of the sun shining and the birds chirping outside your window on the first nice day of spring after a tough winter, when you would love to take the day off and go somewhere to enjoy the bounty of nice weather. It means having "to do" lists and detailed schedules before you. (You can install these in your computer and have the computer ring bells and otherwise alert you to impending schedule requirements.)

Can you handle this all day and every day without fail, and with only carefully calculated deviations or exceptions? If you can, by all means do so; it will add greatly to your probabilities of success. I can't, and I suspect that those who can do so at even 75 percent compliance are relatively rare birds. In fact, they are probably the antithesis of those who seek to be independent entrepreneurs, those of us who rebel at regimented order in our working lives. So let's "get real": There is a difference between idealized or theoretical time management and real-life time management, and it represents a compromise between those extremes. (Bear in mind that those of us who venture into independent, home-based, one- or two-person business ventures as a way of life are among the individuals who are least likely to be willing to regiment our lives in a rigid and constrained formula of management!)

I rather imagine that each of us must devise and develop our own time-management system to fit our own situations. In my case, my office is the room adjoining my bedroom, out of which I tumble at, usually, 6:00 to 7:00 A.M., ready not for work but ready for a morning regimen of coming to full consciousness. (Long gone are the days when I leaped from bed in the morning, fully conscious and ready for whatever the morning would bring. Don't be unduly dismayed if you find that you, too, now need that morning ritual of gradual awakening.)

The first order of business is the morning newspaper, a cup of coffee and the morning crossword puzzle. Then to my office and computer, usually by 8:00 A.M., but sometimes earlier. (I refuse to punch a clock, even symbolically.) At that point, although now awake and coherent, I still need a transition to a working mindset and awareness that it is a business day, so I check my e-mail and other cyberspace connections for my messages to help me achieve that mental/psychological state.

Finally, by about 8:00 to 8:30 A.M., sometimes as late as 9:00 A.M., I am ready for serious work, which will continue until 5:00 to 8:00 P.M., usually, with a brief lunch break, spent at my desk. However, since we maintain a post office box for our business address, I usually go off to the Post Office every morning between 10:30 and 11:00. I then handle my mail and correspondence until they are done, and then go back to my daily work chores until quitting time.

That is a normal weekday schedule, although it may also be a weekend schedule, if deadlines are pressing me or the weather is too depressing to go out and do anything pleasurable.

Fortunately, I don't do much administrative work, other than marketing and correspondence. My spouse handles the bulk of the bookkeeping, dunning the slow payers, and other such chores for us so that I can concentrate on production, but I do have to do some marketing work during the day. So I cut out a few hours for that, when necessary, without precise advance time planning. (I may get an idea for another book at any time, and get to work on it immediately. Or a prospective client may call suddenly and offer some work, but require a bit of negotiation first.)

I suspect that this kind of informal or semiformal work schedule is more typical of most home-based business owners than is the carefully detailed to-do lists and other appurtenances of the bean-counting psychology that afflicts many in our society. Success in home-based business, I think, depends much more on individual instincts and intuition than on formal planning and rigid organization. That is not because the one method is superior to the other, but because most of us are likely to resist more than token conformance with highly disciplined schedules.

While we are on the subject of time planning, let us consider a related problem of schedule conflicts, a too-often occurring headache of custom service businesses in which the client requires that you project a completion date. We tend to be overly optimistic in this too often because we believe that the client expects an optimistic due date. That can lead to related problems, such as conflicting schedules.

Schedule Conflicts

Conflicting schedules are a recurring problem in contracting, regardless of whether you are a large firm or a one-person, independent contractor. You can expect to encounter the feast-or-famine dilemma periodically, and be forced to struggle through it somehow.

The reason such conflicts arise are clear, although the reason for one cause, the feast-or-famine phenomenon, is not always clear. Perhaps it's one of the perverse laws of chance that Murphy commented on so eloquently. However, it is a fact that your most intensive marketing may produce absolutely nothing for a lengthy period, and then suddenly spew forth a flood of riches in the form of several contracts offered at once, each more enthusiastically offered and more urgently needed than the other. After the dry spell you have experienced, with all its worries, and mindful of how much energy and money it has cost you to get these offers, you are most reluctant to refuse or lose even one of the contracts. You therefore ponder ways to keep all by trying to work out schedules that will keep the clients satisfied, while you know in your heart that it will be extremely difficult to meet all those optimistic dates you projected to help you win and keep the contracts.

First Steps in Coping with the Conflicts

The first step is to determine the priority of each contract as the client perceives the urgency of his or her need. To do that, unless you already have some indication in advance—e.g., from an RFP to which you had responded originally or from some other earlier inquiry—you manage somehow to ask each client about the urgency of the project. That will enable you to put the projects in rank order of schedule priority immediately.

A second step is to arrange as long a working schedule (from beginning to deadline, that is) for each project as possible, again as a result of consultation and negotiation with each client.

The Planning or Kickoff Meeting

One excellent way to do both these things with each client is by asking the client to meet with you for initial planning and coordination, where you can carry out your secret agenda of negotiating a favorable schedule. In some cases, this is a first postaward event anticipated and called for by the client in the original contract planning and bid or RFP solicitation. If it is not, you may and should request such an initial meeting.

That is a reasonable request, and clients will generally agree, especially if you are careful to frame the request as a move in the client's interest—e.g., to ensure that you have a complete understanding of the need and the client has a complete understanding of your plans, with opportunity to comment or make suggestions.

In these "kickoff" meetings, you can explore many possible options. Where the client appears to have no special urgency to get the job done, find a reason to schedule the job completion well into some future date. (You might suggest that if the client can wait until you get your decks completely clear, you will be able to deliver some extra benefit, such as introducing into the project some new ideas you have been exploring. (In any case, as in marketing anything, do not ask for a drawn-out schedule on the basis of your need or convenience. Instead, you must show how the client will benefit from a lengthened schedule, if you hope to win agreement and approval.)

As an alternative, suggest a much longer contract period than it might normally take so that you can multiplex your work and get work going on more than one project concurrently. In some ways, that actually makes you more efficient. Working a half-day each on two projects is often less tiring than spending a full day on one project: The midday change to a different challenge often freshens the mind. In any case, a lengthened schedule definitely eases the problem considerably, giving you some slack.

There is also always the possibility of subcontracting some of the work to a professional associate or other vendor when you know that you cannot meet the deadline without help and it is urgent that you do meet that date. (I have found that to be an excellent idea in many cases where my schedules were in conflict.)

Finally, there is the desperate option of agreeing to an impossible schedule and plowing into the jobs, fingers crossed, hoping you can come up with a good story and your clients will understand when you have to postpone the delivery date later. That is the least desirable way to handle schedule conflicts. If you are unable to get the needed slack in preliminary negotiations, as suggested here, at least do not wait until the deadline is at hand before you talk to the client about your need for more time. Raising that point in midproject is usually easier on the client, giving the client some slack to alter his or her own plans accordingly.

The Other Side of the Coin: The Vendor Problem

The problems you have are not different from the problems others have, even your own suppliers and subcontractors. They, too, often have schedule problems, sometimes adding to your problems. Good management dictates that you prepare for this as a contingency.

No matter how self-sufficient we make ourselves as independent contractors, there are some needs for which we must depend on vendors. Printing is one service for which we usually must depend on a vendor, for example, but there are many others, such as delivery of promised software, processing of slides or photographs, report of laboratory tests, completion of drawings or construction of a prototype, to name only a few possibilities.

Your own schedule may be endangered by a vendor who accepted your contract or subcontract, but finds his or her shop overloaded and wants an extension you cannot grant. I once made arrangements in advance with a printer to print 250 copies of a thick proposal that was most important to us and had to be delivered on time. (Usually, a proposal has to meet a most firm deadline, and is a totally worthless document and total waste of time and money if it misses that deadline even by minutes.)

At the eleventh hour, as we were about to deliver the copy for printing, the printer announced that he was overloaded with work and could not deliver the printing to us on time.

My response was that he could forget about the job. I would take it elsewhere. That caused him considerable dismay. How could I do that? If he could not meet my schedule, what made me think anyone else could do so?

I said (with apparent great confidence) that that was my problem, not his, and I would attend to solving it. At that, he caved in, swore he would push other work aside and do our job on time. Ultimately, he did just that, afraid to stall any longer. He did so because he was convinced that I would, indeed, pull the job away from him and get it done elsewhere. And I would have. I had several other hungry printers ready to take on the job.

The lesson is this: Whenever you need an important task or function performed over which you do not have direct control, have more than one way ready to get it done. Have a nominal source, but always have at least one alternative to make you independent of that prime source, if necessary. (I call this my menu plan, with choices from two or more columns available.) Again and again, I have found this to be wise precaution.

You must take the *independent* adjective quite seriously and work hard at being independent in every sense of the word. This is one of the ways to maintain that independence.

Zoning Problems

Owners of home-based businesses sometimes run into problems with the authorities of their communities, especially zoning commissions and boards. It is the responsibility of such commissions to see to it that residential neighborhoods are not turned into commercial or industrial neighborhoods by permitting commercial or industrial uses of residential properties. To put it plainly, it is against the law in most jurisdictions to conduct a business enterprise in your private home.

Nonenforcement Is Commonplace

Some jurisdictions are liberalizing their zoning ordinances, but many simply prefer to refrain from enforcing them when the home-based business is not a nuisance and does not reflect a change in the fundamental residential nature of the neighborhood. Thus in practice, unless someone complains, the authorities will never trouble you. Such complaints may come from neighbors, if what you do constitutes a nuisance to them. That is, if there are industrial noises emanating from your home, if trucks frequently pull up in front of your house to pick up and deliver packages, if you have a stream of customers coming and going all day or are otherwise highly visible as a business activity, if you even have discreet signs exhibited on your property, or otherwise openly advertise the fact that you conduct a business in your home. Then again, even if you are entirely and totally discreet, neighbors or even others may complain out of spite.

Once a complaint is made, the authorities are forced to investigate it, and if they substantiate that what you are doing in your home constitutes an illegal business activity for a residential property, you will be required to cease. In fact, you may well have to appear in court, charged with a misdemeanor. I once kept an electronics shop in my garage, and although there was no noise, no sign, no advertising and no customers or deliveries coming or going, a neighbor complained for some reason (I never learned why). I was summoned to court. The judge was sympathetic and advised me that there was no choice but to require me to move my shop and charged me $9 court costs (this was a *long* time ago), but applied no penalty.

Legality versus Practicality

This is not to encourage you to break the law, but zoning laws are antiquated in most places and difficult to change. The authorities, recognizing the obsolescence of the zoning laws in today's world but unable to change the laws easily, often prefer to ignore the minor violations of such quiet home-based businesses. Writers, artists, real estate and insurance brokers and many others have operated offices in their homes for many years without arousing the authorities. You can, however, ask for a variance, an exception that authorizes you to operate a business in your home, despite the ordinances. (The zoning commission may decide that what you propose to do is not a violation of their laws.) That is a hazardous step because it is likely that your request will be denied, and then the fat is in the fire: You will have to move your business out of your home. If you decide to try this approach, proceed with great caution. First, check out the local zoning laws; if you think you can make a good case for a variance, consult a lawyer before you take the next step.

Other than that, there are several steps you can take to minimize the possibility that anyone will even suspect that you keep an office in your home, much less have occasion to consider that what you do constitutes a nuisance:

- Use a post office box as your business address. Many people use the alternative of one of the many store-front services that have sprung up in recent years to support small home-based businesses as a mail-drop, local address for sending and receiving packages and other conveniences, such as copying and faxing.
- Receive clients in your own office only by appointment, and make that as infrequent as possible. Try to meet clients in their offices or over lunch, whenever possible.

- Be closed-mouthed about what you do and, especially, where you do it. No matter how friendly others may be, you cannot predict how they will react to the information that you are self-employed and keep an office in your home.
- If your business involves activities that are noisy or would otherwise attract attention, consider having those activities done elsewhere for you by a vendor or subcontractor.

Cash Flow Problems

The inadequacy of one's cash flow is a common problem in businesses of all sizes. Even the business with ample sales and profitability can easily experience the problem of "having too much money on the street" (owed to you), with the result of not having enough cash on hand to meet expenses.

This comes about when you must invoice your clients and wait for payment (probably for 30 days or longer), but without enough operating capital to finance operations—to pay your bills and draw a salary, while you wait for clients to pay.

You may or may not find that you have this problem, depending on how much operating capital you have to begin with and on the nature of your work. That is, it depends in part on whether your work is such that you can normally demand to be paid immediately for your work. If you normally handle small contract jobs for private citizens, you can usually require payment promptly on completion, and possibly get some of your money in advance. If your work is of fairly large contracts and especially with large organizations, you probably will be expected to bill your clients and wait for payment. Unfortunately, companies rarely pay in less than 30 days nowadays, and often take longer than that. You may thus soon develop a problem of too many receivables and not enough cash in hand. When you have a great many accounts outstanding in that manner, there is a great possibility that some of them will turn out to be bad accounts, accounts that you will be unable to collect.

You Should Not Be Forced To Finance Your Clients

If you find yourself in the position of extending credit to everyone and waiting for payment a long time after the job is finished, you are actually financing your clients: Until you get paid, the client's needs are being satisfied at your expense—with your money. Even if you have enough capital to squeeze by while doing this, you ought not to have to risk your business survival in this way. Even the largest corporations do not permit unrestricted drain on their operating capital. Taken to the extreme—if your accounts receivable becomes so large that you cannot handle it out of your own capital—this kind of problem can make it most difficult, even impossible, for you to remain in business. Remember that you are extending credit when you undertake a project and see no money up front. Extending credit always involves risk, something you must minimize. There are remedies, but they require you to be completely businesslike and act defensively.

Deposits and Retainers

As a consultant and independent contractor, I soon learned to always insist on a retainer or deposit up front, with only certain carefully qualified exceptions (more on that shortly). My usual practice was to ask for approximately one-third of the estimated total price as a retainer. I asked for a second third of the total at some agreed-upon midpoint, such as the client's acceptance of a draft or detailed outline and working plan. The final third was to be paid on completion.

Not only does this help your cash flow, but it helps ensure you against loss. Experience showed me that when I did not get a retainer (as in my earliest and most inexperienced days), I was in great danger of never getting anything.

When I did get the retainer, I had no problems collecting the remainder of my bill. There was only one exception to this, but I had already collected more than 75 percent of my bill beforehand, and so was not too badly hurt. Except for that one case, the client's reaction to my request for a retainer and midproject payment were good indicators of my client's sincerity and ability to pay. I made this a firm policy when I learned that making exceptions to this, except as I shall note here, entailed more risk than it was sensible to accept, a risk that cost me a great deal more money than I could afford to lose.

Progress Payments

The exceptions I would make to my demand for retainer were in contracts with federal government agencies, where it is somewhat difficult for the agency to arrange advance payments, and with large corporations when it would take several weeks for the paperwork to be processed, and the job had to be started at once. In those cases, I was not concerned about the client's capability to pay the bill, and there was usually no difficulty in getting progress payments—e.g., every two weeks—if the job was to be fairly lengthy. In fact, however, sometimes even the supercorporation could and would arrange to get me a retainer or up-front deposit. Certainly, it does no harm to ask for it.

Minimizing Travel Expenses

Accepting out-of-town jobs entails special expenses. There is airfare, of course, but there is also the expense of local travel to and from airports and perhaps of other travel while working on the assignment. There is the daily expense of lodging and meals also. This, too, can amount to a tidy sum, and what is worse, there is no profit in this for you; it's all pure cost.

I found that in most cases, it was possible to persuade the client to handle all this cost, except for a few minor items, in a way that made it billable directly to them. The client can arrange to send you tickets or have them ready and waiting at the airline's check-in area in your airport. Clients also can arrange to have your hotel charges and even the cost of a car rental billed to them. I never have found it at all difficult to have a client agree to this, making it a business expense you do not have to bear even temporarily.

Selling Your Paper

"Selling your paper" or "discounting your paper" is business jargon for borrowing money against your receivable accounts. (It is also known as "assigning your receivables," which is the legal term for selling or otherwise transferring them to another party.) You may have experienced this in buying a house or an automobile. Sometime later, after you have made settlement on the property, you got a letter and perhaps a payment book from some bank or other lending institution you knew nothing of, advising you that you would make your mortgage payments to that institution.

The bank had taken over the loan from the mortgage company you dealt with at first, one of the ways a bank finds loans to make so that it can earn money. The mortgage company had sold your note ("assigned" it), at some discount, to the bank. If the note or mortgage was for $100,000, for example, and the bank bought it for $92,000, they stood to realize $8,000 in gross profit.

When you buy any appliance or other item from a dealer and finance the purchase, the same thing happens. The dealer discounts the note with a bank, getting the money represented by the note, less some discount that represents the profit the bank will realize.

That isn't easy for you to do as an independent contractor. Banks prefer to buy notes covering property that acts as collateral, at least in part, for the money paid, and bankers seem to have a great deal of difficulty understanding the concept of "labor inventory" as an asset. Thus, if the assignment of an invoice was for service, the banker may be a bit uneasy and want the assignment to be "with recourse"—the dealer guarantees the payment to the bank. The bank will want some assurance that the dealer is capable of making good on such a guarantee. If it is without recourse, and the bank accepts the risk, the bank is more particular about the soundness of the receivable and probably wants a relatively large discount to compensate for the risk.

Although you probably will not find it easy to sell your modest amount of paper to a bank, there is an alternative: It is possible, in many cases, to do business with a "factor," "factoring" the note. This is the same thing, but the factor is usually an individual with money to invest. The factor will require a much larger discount than the bank does, and probably will accept only those notes on which you will be responsible for payment. You will still undertake the risk that always accompanies sales on credit.

Danger Signs

Although this section addresses cash-flow problems only indirectly, this is a good place to alert you to a danger sign or two concerning money and getting paid for your service. You will recall the admonition to never volunteer your price, but always to wait for the prospect to ask you what the price will be. There are at least three signs of impending hazard with regard to price or cost:

1. The prospect who starts by asking the price first.
2. The prospect who never asks the price.
3. The prospect who says that money is no object.

An example of the prospect who seems to be concerned with price first and foremost was given earlier. The person seeking "the cheapest guy in town," is almost guaranteed to become a bad contracting experience if you come to terms with him or her and consummate a contract.

I would terminate any inquiry that starts with price as quickly as possible. (There may be an exception, but I would be pessimistic about expecting to encounter an exception to this.)

The prospect who never asks the price either is not a serious prospect to begin with; has no intention of paying you, if you can be persuaded to undertake the project on credit; or hopes vaguely to cross that bridge later, which also means that it is unlikely that you will ever be paid. In any case, it is wise to be highly apprehensive when a prospect appears to be uninterested in the cost of what you are being asked do. At the least, that prospect is probably not sincere, not really interested, but is fishing about for information, possibly just sizing you up for a competitor. The sooner you terminate that interview, the better for you.

There is a variant on that prospect. If you are a marketing consultant of some kind, being asked to help a prospect develop a proposal, a direct-mail package or some other kind of sales campaign, you may encounter the prospect who wants to give you a share of the anticipated profits or a piece of the contract. I yielded to this kind of argument a few times and lost every time, even when my service helped the client to achieve success in whatever marketing project he or she was undertaking. The reasons they can find for reneging on the deal are infinite in number, and even if you manage to write a powerful, binding contract with them, you will be dismayed to learn that they are judgment-proof. So was my unhappy experience.

The prospect who says quite openly that cost is of no concern—"money is no object"—is in the same category as the prospect who never asks the cost. I would waste no more time on a prospect who says that money is no object. That is a client who will never pay you. That's why money is no object: The prospect doesn't have any, and wouldn't pay you even if he or she did.

An easy way to qualify the prospect and terminate these time-wasting sessions is to introduce the requirement of an advance retainer. That usually confirms quickly that your prospect can't or won't pay you "earnest money," and enthusiasm wanes quickly. That leaves you free to go on to more fruitful meetings.

10

Writing Needs and Skills

Despite the high level of literacy in modern society, many of us still resist learning how to write well, despite its importance to our careers and businesses.

The Need To Write Is Inevitable

We live in the information age, and whether information is collected, organized, stored or propagated and presented by wire, computer screen or ink on paper, it first has to be written. The bulk of the writing required by and carried out for the business and professional world is done by those of us in the business and professional fields of activity, by independent contractors as much as by others. It is done by us because it must be done by us, although some of us choose to be freelance writers working in highly specialized and often technological areas. Still, many freelance writers cannot or by preference do not handle the more specialized areas.

The Many Things That Need To Be Written

One area in which many of us, executives, professionals, business owners and others in general, tend to be deficient is in writing. We tend to underestimate its importance, probably because so many of us dislike writing, for whatever reasons. (I confess that I really have never understood the common dislike of writing, having been myself blessed with an affinity, as well as an enjoyment for, the practice of writing.) Whatever the reason for the common reluctance to write, the ability to write clearly is not terribly difficult to master, and it pays off in many ways, whatever your field as an independent contractor may be. At any time you may be called upon to help a client develop an annual report, a speech or an article for a trade or professional journal. You may invited to contribute to a newsletter, affording you some free publicity that may help produce new business. You may be called upon to develop a manual or other user instructions in the course of one of your projects, for example, or, if you are disposed to it, you may wish to undertake a contract to develop a manual or some other sort of user documentation. That is itself a burgeoning field of contracting and subcontracting today as a major source of work for freelance technical writers. But there are also reports of various kinds, especially if you are doing any kind of pioneering or research work. Here are some of the items you may be called upon to write or help a client write:

- Proposals
- Product or service brochures
- Press releases
- Reports
- Specification sheets
- Advertisements
- Journal articles
- Direct mail packages
- Speeches
- Newsletters
- Sales letters
- General correspondence
- Capability brochures
- Magazine articles
- Professional papers
- Books

We covered some of these items in earlier chapters, but not all the more important ones to your career and business as an independent contractor. It is not possible to cover the entire spectrum of business writing here, but there are a few of special interest and potential value—representing contract opportunities and profitable diversification—that we can touch on.

Reports

There are probably as many kinds of reports written in business operations as there are kinds of brochures, perhaps even more. Most are purely for information and reporting or documentation purposes, while some others tend to be almost as much for advertising and promotion as for other purposes.

Annual Reports Most corporations produce annual reports; corporations with stockholders must produce such reports. In small corporations such a publication may be little more than a few typed sheets bound with a staple in the upper left-hand corner, and presenting the factual (or factual-appearing) data that accounts for what the corporation has done in the previous year. Of special interest to stockholders are the several accounting reports, showing such important matters as profit and loss, proprietorship and other such things. Characteristically, however, annual reports are expensive, slick publications in which corporate managers explain to stockholders how well their corporation has been managed during the year just elapsed, what investments have been made, what great plans exist for the future and why the red ink is really not red ink at all, but is something that will pay out in future years.

In short, an annual report is one in which the corporation's managers—the officers of the corporation, that is, especially the president, the chairman of the board and the other top executives—attempt to persuade the stockholders that they have been doing a fine job of running the corporation, ought to be kept in office and should be supported and encouraged to go on doing what they have been doing and plan to do in the future.

Quite often, the managers have bad news for the stockholders: losses instead of profits, declining sales, white-elephant divisions that have to be dumped, expensive projects that have gone sour, the need to bypass the payment of dividends this year and other distasteful items. The need for persuasion in such circumstances, is acute, especially since the numbers that glare so starkly from the glossy pages have been prepared by certified public accountants who must, in fact, certify them and swear to them. (The numbers can be made somewhat more palatable to the uninitiated by certain euphemistic devices, but the knowledgeable stockholder knows how to read an annual report and is not easily deceived by euphemisms and other persuasive stratagems.)

Progress Reports Projects and other ongoing efforts, especially when they are special custom efforts or are long-term efforts, generally require that progress reports be prepared and submitted on some regular basis such as monthly or quarterly. The recipient of the report wants to be assured that satisfactory progress is being made, all problems encountered are being solved, and that the budget and schedule are intact. Too, in the event that problems have shown up, the recipient wants to be kept informed about progress in eliminating or solving the problems.

The writer of the report has analogous goals, of course, and wants to assure the recipient, who may be a superior or a customer, that all is well, all is being coped with satisfactorily and all will end well.

Obviously, as in the case of an annual report, the reporter tries to soften bad news as much as possible, and at the same time tries to paint his or her own work and abilities in the best light. However, there may be other objectives, given different circumstances in each case. In one case, the series of progress reports may be calculated to prepare a client or superior for final results that are not those originally expected. A series of a dozen or more monthly progress reports may be presented so that the final report will prove acceptable. That is, it may be necessary to break bad news gradually, over many months, to persuade the customer or superior to accept it. In any case, progress reports and final reports on projects are usually inherently persuasive in nature.

Financial Reports Some of the routine financial reports prepared by accountants are those called "P&L"—profit and loss—statements and balance sheets. Many other types of financial reports can be and are prepared to reflect the condition of the enterprise and assist managers in making wise decisions. Some of these reports serve only the purpose of helping managers perceive factors that should be considered in making decisions or that will guide wise decision making. Others serve this purpose, to some degree, but are principally final reports on the overall condition of the enterprise and are prepared for owners as feedback on how the business is doing. (For example, this is the kind of information that appears in the corporation's annual reports.)

There is a problem with such reports in interpreting what the figures mean, in terms of their overall significance. For example, suppose a P&L reveals that the company lost money again, for the third year in a row. But suppose that another kind of report showed that the loss was not an actual operating loss on the year's business, but was caused by residual effects of the previous year's problems—that is, that the business now is turning an operating profit and will show an actual profit as soon as last year's disaster has been completely liquidated. This means that it is sometimes necessary to generate reports to explain and understand the significance of other reports. Profit by itself, for example, may be less significant than return on investment (ROI), so it i essential that ROI reports be generated.

Financial reports also often are needed to bolster managers' arguments, when they are urging some course of action and meeting with opposition in their companies. A manager trying to win approval from the board of directors, for example, must present some pretty strong arguments, especially if the recommendations are going to require substantial investments or entail risks. Financial reports are almost invariably a must in such situations, if the manager urging the action is to have any hope of prevailing.

Research Reports Anyone engaged in research and development of any kind must produce research reports. These are, in fact, progress reports, and if they are reporting on projects that are supposed to produce a useful end product of some kind, they need to be persuasive and convince the reader that the need for continued support is great indeed. On the other hand, if the effort is pure research, the report must persuade the reader that the research is being conducted logically, methodically, in accordance with sound principles of research practice and that results are being sensibly evaluated and analyzed. The writer is inevitably tempted to present himself or herself and the work in the best possible light, even when no particular end result is demanded, other than information or knowledge. In many respects, research and study reports are very much the same as progress reports. (Study reports are almost identical to research reports, for all practical purposes, except that they may cover surveys instead of laboratory work.)

Professional Journal Articles

Those in the professions usually find it desirable to publish in professional journals. Such journals are dedicated to the advancement of knowledge in, and status of, their fields—medicine, dentistry, engineering and others—and encourage scholarly articles of interest to members of their professional societies, who read the journals. For the individual who writes such articles or papers and has them published, the achievement is one that adds to the individual's prestige and personal reputation in the professional field. You may wish to publish in such journals to enhance your own image and promote your contracting business, but it is not uncommon for clients to call on contractors for help in getting their own articles and papers written and published.

Such journals are circulated among one's professional peers, of course, since they typically are published by professional societies, and are read by one's peers. The objective in writing such articles, then, may be a sense of professional duty and pride, but it also may be, and often is, ego gratification, since publishing in professional journals is generally recognized as an accomplishment and therefore is prestigious. To realize the accomplishment, the author of such an article must generally manage to persuade the editors of the journal that the article is scholarly and highly professional, and presents information of great value not otherwise known in the profession. The result of this is a writing style, which is itself somewhat ponderous and jargon-ridden, but presumably makes its own contribution to persuasion. Always make sure to consult the journal's "Author Guidelines," usually published in the first issue in each volume, for the journal's style guidelines.

Trade Journal Articles

Every field, professional and other, has its share of trade journals— magazines, usually, but sometimes tabloids, too. TV repair shops may subscribe to *Electronic Technician*, for example, while a department store operator may subscribe to *Sew Business,* and a radio station owner may subscribe to *Broadcast.* Such journals ordinarily are published by private, for-profit publishers, and have a somewhat limited circulation, as compared with magazines appearing on the newsstands, and they usually carry a great deal of advertising, as do newsstand periodicals.

In some fields, such journals are "highly professional" in that they carry highly technical articles, often quite abstract and having usefulness to a quite limited readership. In other fields, such journals are quite broad in their appeal, and even a lay person would have no difficulty in understanding the articles (although the lay person may not be at all interested!).

There is a certain amount of prestige in being published in some of these, especially those that closely resemble the professional journals. In some cases, such journals may be read by people with whom you might do business, and having your article published there or getting your press releases used there may represent useful advertising for the writer. Therefore, trade journal articles may have to be persuasive to carry out their missions.

Popular Magazine Articles

Much of what has been said for trade journals will apply to some popular magazines. For example, it may be useful for the computer repair shop owner to write articles for such newsstand magazines as *Popular Electronics* or *Byte*. Of course, the same thing applies to press releases published in popular magazines of general interest. For many, this represents free advertising, which is even more important and more valuable than that achieved with trade journals because the popular magazines normally have a much wider circulation.

Speeches

One of the most effective ways to get publicity for certain callings (such as consulting and other professions wherein commercial advertising is impractical, for one reason or another), is by making speeches before various groups. Writing a speech is a special problem in writing, and it is best to write your own speeches, if at all possible. One reason for this is that the speech you write for someone else may sound fine when you deliver it, but may either trip up someone else's tongue constantly or may simply not fit the other person's speaking style. Many well-written speeches are discarded simply because the individual for whom they are written is not comfortable with them.

Opportunities to make speeches exist at every meeting of any organization whose existence has relevance for what you do or offer, at national conventions and conferences, at seminars, at trade shows and at many other gatherings, social and business. "Relevance" does not mean that the group is one of people in your own field. You may have things to say that are of interest to almost everyone, although you may have to "slant" your presentation differently for different audiences. One man I knew a few years ago had a speech he used for years about the evolution of modern educational technology, and he presented it to a great many different kinds of groups, including writers, editors, educators, professional specialists and sundry others, managing always to relate the message to his listeners' interests without making substantial changes in his address, simply by how he introduced his subject.

Persuasiveness in public speaking has certain, special considerations, which we'll probe when we discuss the subject in greater depth later.

Direct Mail Packages

The sales letter is a centerpiece of the direct mail package, as a rule, although often eclipsed by the weight of other enclosures and inserts in the package. Typically, a direct mail package includes the sales letter, a brochure or broadside of some sort, an order form and a response envelope as the obligatory minimum. Frequently the package includes even more—sometimes as many as a half-dozen different items in a kind of scattergun approach, apparently in the belief that if enough arguments and persuaders are fired in the direction of the addressee, something ought to work. (That belief is sometimes expressed in a little epigram, "The more you tell, the more you sell.") And if that isn't enough, many direct mail packages also are sent off in a large and thick envelope, the outside of which is covered with various other brief but highly charged messages, often in two or three colors.

Writing User Information

The word *manual* tends to conjure up an image of a bound book of some sort, and user instructions are in fact often voluminous enough to require that they be presented in bound volumes, even in multivolume sets, when they are developed for the users of large systems. At the other extreme, there are the user instructions that are on a single sheet of paper—even contained among other information on a label or wrapping—guiding the user in a simple task, such as assembling a plastic flower-pot stand or fastening a simple appliance to the wall. The difference between these is of scale or scope, but not of kind. For our purposes, *user manuals* will refer to all and any kind of user instruction, regardless of the size or nature of the physical product. Most of the information and suggestions will be in the reference frame of large bound manuals, but obviously much of the recommendations made here are overblown for many uses, and the presumption is that you will adapt the recommendations to your specific needs, even for the writing of a one-page "manual."

I have used two terms here: *user manual* and *user instruction*. You may encounter a third term, *documentation*. That term is much broader in scope and includes user instructions or manuals among written items, such as reports, contract specifications and records in other forms than on paper, such as computer tapes and disks. (In many cases, user manuals also appear in that form.)

One more item requires clarification: Typically, user manuals are conceived as being written about physical products. That is not always the case. More and more, in today's increasingly service-oriented economy, user manuals for intangibles—services such as access to online databases, use of long-distance telephone services and operation of computer software—are necessary. What is written in these pages applies equally to user manuals about both physical products and services.

Who Is Responsible for Understanding the Instructions?

It's true enough that there are some individuals who never "get the word," as some people in the U.S. Navy say, no matter what pains you go to in trying to make understanding possible and misunderstanding impossible. But that is not a proper excuse for failing to make all reasonable efforts to make it easier to understand than to misunderstand.

It is important to get priorities straight, and the first one is that of where and on whom the prime responsibility for understanding falls: It falls on the writer, not on the reader. For example, an editor in a technical manuals project once called me to task for my use of the word *epitome*, where I had remarked that the arithmetic unit of the computer is the epitome of computer operation generally. I was puzzled at first by the complaint because I was certain that I had used that word quite correctly. And then the suspicion came to me gradually that probably the editor himself did not understand the precise definition of that term—it does have a subtle shading—and thus did not understand its proper application at all. I persuaded the editor to look the word up, received his embarrassed apology and returned to work. But I was still troubled with a vague unease about the matter, although I was not sure what it was that so troubled me. But I soon reasoned out what was still bothering me about it when I turned to several other writers whose opinions and abilities I respected and asked them to define the word for me. That crystallized my problem: not one of them knew the precise definition—they all had only vague ideas as to its approximate meaning—and could not, therefore, have used that word appropriately, much less judged whether I had done so.

I immediately, and most regretfully edited that word out of my copy and found another term to use, a synonym not quite as "right" for the purpose but one that most readers would probably understand. For my checking with experienced writers had demonstrated clearly to me that few people would be likely to have a full and accurate understanding of that word if so few professional writers knew the word well enough to use it.

The point is, of course, that being right is not important and is certainly not the issue; being understood is the issue, and it is what must have the first priority.

A Basic Fallacy

One obstacle impeding the development of better user manuals is the very fact that we tend to refer to *writing* user manuals, as though the ability to use the language well in writing is the key necessity in producing good user manuals. Unfortunately, that is not the case at all; writing as an art itself is only a portion of the task, not more than one-half at best, and not the more important portion either. I have more than one well-written manual explaining computer software programs I own, but I still have great difficulty in solving an occasional hardware or software problem because the author of the manual failed to anticipate my need for certain details. The author's writing was not at fault, but the author's understanding of my needs, as a reader, was very much at fault. The author lacked or appeared to lack that most important quality of a writer, *empathy*.

Empathy

Empathy is the understanding of another's situation, feelings, motives and other outlooks. In this case, it is understanding of a reader's interests and needs. My quarrel with most computer software manuals is their shortcomings in anticipating so many of my needs as a user who is not a computer expert.

Preparation for Writing

Is it the author's responsibility to anticipate my problems? It certainly is. The purpose of the manual is to guide me in using the hardware or software about which the manual is written, and that includes steering me through the shoals. The problem is a rush to the pen (an allegorical pen, of course), a rush to begin writing before planning in detail, apparently before planning at all in some cases, and perhaps even without truly critical review and revision—in short, without proper preparation to begin writing. It is as though you set out to build a house without plans, improvising the design along the way with whatever ideas occurred as you proceeded!

It is this lack of preparation to write that makes writing so difficult for many. Without proper planning, the problem is not one of how to write well, but of what to write about—what subjects to present and what to say about them. (Of course, the converse is true: with proper preparation, writing becomes much easier.)

Therefore, before moving on with specific information and suggestions for creating the various kinds of user manuals, this brief introduction to and suggested guide to planning and preparation is offered.

Key Planning Events To Be Defined in Advance There is no mystery about the key elements that must be defined before beginning to write a user manual, regardless of its size or scope. The author(s) must know the purpose of the manual (assembly, installation, maintenance and so on) and the identity of the user or reader for whom it is intended. In fact, it is a practice in well-organized technical publications groups to develop an actual profile of the reader, defining the reader's general knowledge of the subject, reading abilities and duties or job description relevant to the manual—e.g., radar technician, computer operator, machinist and so on.

Needs Assessment That is not all that needs definition, of course. In fact, having identified the general goal or purpose of a given manual, that identification must be further developed into a detailed definition of what the user needs from the item, and here the characteristics of the item itself must be taken into account. That means that planning must include research in an iterative mode because data uncovered in research often result in reworking of the original plan. Research into the characteristics of the item should result in feedback to the outline of content.

A First Step A generalized outline is offered here to identify, in general terms, the possible items to be considered in planning and preparation. A more detailed outline will be developed from this after a brief discussion of the preliminary outline and its use.

I. Identify characteristics and needs

    A. Item characteristics
        1. Technical
        2. Nontechnical

  B. Item Needs
    1. Assembly
    2. Installation
    3. Operating procedures
    4. Maintenance

  C. User characteristics
    1. Lay users
    2. Qualified users

  D. User needs
    1. Assembly instructions
    2. Operating instructions
    3. Maintenance instructions
    4. Technical/functional descriptions
    5. Special training

 II. Sort and merge

  A. Correlate and combine characteristics and needs

  B. Organize into order of priority

 III. Define major chapters or manuals

 IV. Develop preliminary (general) outlines for each

 V. Estimate size of each

 VI. Draft preliminary design

Developing Preliminary Data You can develop preliminary data before the first planning meeting or as a result of and follow-up to that first meeting. At the minimum, the plan for doing this ought to be presented at the first meeting, and the business of that first meeting ought to be to discuss this and select the team to work on it. It is also possible to do some of this—probably the first three major items—in advance of the meeting.

 The outline reveals the basic strategy and the general logic of the proposed methodology for developing a total needs profile and thus providing a sound basis for design of the documentation and definition of the necessary content. It is actually a working blueprint as it stands, but it still calls for a great deal of estimating based on personal judgments, and the goal here is to make the process as objective and methodical as possible. The following expansion of the original, general outline, while still calling for individual judgment and estimates, adds benchmarks or reference points to aid you in making those judgments and so furthering progress toward objectivity and method. (You are, of course, to use this as a model or checklist and adapt or improve on it as necessary and appropriate.)

I. Identify characteristics and needs
  A. Item characteristics
    1. Technical
      a. Complex or state-of-the-art technology
      b. Classic, medium complexity
      c. Moderately complex
      d. Conventional technology, not complex
      e. Involves single technology
      f. Spans two or more technologies
    2. Nontechnical
      a. Complexity of many elements, parts, applications and so on
      b. Medium complexity of elements, parts, applications and so on
      c. Simple, not complex
  B. Item Needs
    1. Assembly
      a. Complex assembly, requires many tools or special skills
      b. Moderately difficult assembly, requires no special tools or skills
      c. Simple assembly, requires only simple hand tools and skills
    2. Installation
      a. Complex, technological installation problems
      b. Moderately complex technological installation requirements
      c. Simple, nontechnical installation requirements
    3. Operating procedures
      a. Technological and complex, involves specialized procedures
      b. Moderately technical procedures

        c. Simple, nontechnical procedures
    4. Maintenance
        a. Preventative
           1. Many routine, periodic procedures
           2. Requires technical knowledge
           3. No special knowledge required
        b. Corrective
           1. Complex troubleshooting required
           2. Knowledge and use of special tools
           3. Conventional methods and tools

C. User characteristics
    1. Lay users
        a. Novices
        b. Expert hobbyists
    2. Qualified users
        a. Highly trained experts or similar items
        b. Highly trained experts, both or all technologies
        c. Highly trained experts, single or main technology
        d. Moderately trained technician or similar items
        e. Moderately trained technician, both or all technologies
        f. Moderately trained technician, single or main technology

D. User needs
    1. Assembly instructions
        a. General guidance
        b. Detailed (cookbook) procedures
        c. Special training
    2. Operating instructions
        a. General guidance
        b. Detailed (cookbook) procedures
        c. Special training
    3. Maintenance instructions
        a. General guidance
        b. Detailed (cookbook) procedures
        c. Special training
    4. Technical or functional descriptions

II. Sort and Merge

A. Correlate and combine characteristics and needs

      1. Match item needs with user needs
      2. Develop single set of needs

  B. Organize into order of priority
      1. Estimate which needs most critical
      2. Sort in descending order of priority

III. Define major chapters or manuals

  A. Assembly

  B. Installation

  C. Functional theory

  D. Preventative maintenance

  E. Corrective maintenance

  F. Troubleshooting

  G. Overhaul and repair

  H. Parts list

  I. Training

IV. Develop preliminary (general) outlines for each

V. Estimate size of each

  A. Number of text pages

  B. Number, sizes of illustrations

  C. Number, sizes of tables

VI. Draft preliminary design

  A. Single volume

  B. Multivolume set

  C. Formats

You can, of course, reduce the scope of this outline or carry it to even more detailed levels, according to the individual situation and need. You also can use standard worksheets, such as the examples shown in

Figures 10.1 through 10.3 to help you anticipate the needs. Note that the worksheets address the three specific areas of interest:

1. The item itself
2. The user
3. The documentation

Writing Tools

All writers need tools to do their job, just as anyone practicing a profession or trade does. So you must learn what the tools are and how and where to use them. But it is easy enough to miss the obvious tools.

Some years ago, as manager of an organization devoted to writing manuals and training programs, among other products and services, I discovered a practical little test that I found to be most helpful in hiring writers for a fast-growing staff. I first screened out those applicants who appeared to meet none of my predetermined qualifications. I then asked those applicants whose résumés made them appear to be suitably qualified to present lucid explanations and instructions in writing to take the test. The test asked the individual to explain, in not more than 100 words, how to write a personal check. The text was to be to a reader specified as someone who was completely unfamiliar with the entire concept of a check, so that the writer could take absolutely nothing for granted but had to explain every detail of the process. Remember, in reading the following, that each of these candidates was a writer with some degree of professional experience in writing; none was a student or beginner.

The results of this little test always were interesting. Most applicants found it difficult to cover the subject adequately in so few words unless they started from the premise that the reader at least knew what a check was. But every now and then one of the applicants managed to do the job adequately with words to spare—using as few as 50 or 60 words only!

How did these latter writers finesse the problem so easily? Quite simply: They proved to have a true understanding of their mission as writers; they understood that their true mission was to *communicate*, not to *write*, using all the tools available to the imaginative writer, whereas the other applicants all thought of themselves as "wordsmiths" only, individuals who knew and could use a wide variety of words and therefore restricted themselves to that single arsenal: words. The few—

FIGURE 10.1 Item Characteristics and Needs Worksheet

The Item Characteristics

Name:_____ ❑ Technical ❑ Nontechnical

Brief description: _____

Technology/technologies: _____

Degree of complexity: _____

Normally delivered to user ❑ Assembled ❑ Unassembled

NEEDS

Difficulty of assembly ❑ High ❑ Moderate ❑ Simple

Requires special tools ❑ Yes ❑ No

Requires installation ❑ Yes ❑ No Requires adjustment ❑ Yes ❑ No

Complexity of installation ❑ High ❑ Moderate ❑ Simple

Requires special tools ❑ Yes ❑ No

Requires operating procedures ❑ Yes ❑ No

Complexity of operating procedures ❑ High ❑ Moderate ❑ Simple

Requires preventative maintenance ❑ Yes ❑ No

Requires corrective maintenance ❑ Yes ❑ No

Complexity of maintenance ❑ High ❑ Moderate ❑ Simple

Requires special tools ❑ Yes ❑ No

Requires detailed functional theory information ❑ Yes ❑ No

Notes: _____

Instructions: Cross out items that do not apply. Use notes section for additional information.

FIGURE 10.2 User Characteristics and Needs Worksheet

The User
Characteristics

General description or job title if appropriate: _____

_____

❏ Lay person ❏ Novice ❏ Expert

❏ Qualified specialist ❏ Highly trained ❏ Moderately trained

❏ In general/main technology ❏ In related technology
  ❏ In similar item

Needs from Item

❏ Assembly instructions: ❏ General ❏ Highly detailed ❏ Training

❏ Installation instructions: ❏ General ❏ Highly detailed ❏ Training

❏ Operating instructions: ❏ General ❏ Highly detailed ❏ Training

❏ Functional theory: ❏ General ❏ Highly detailed ❏ Training

❏ Preventative maintenance: ❏ General ❏ Highly detailed ❏ Training

❏ Corrective maintenance: ❏ General ❏ Highly detailed ❏ Training

❏ Overhaul and repair: ❏ General ❏ Highly detailed ❏ Training

Notes: _____

_____

_____

_____

_____

_____

Instructions: Cross out items that do not apply. Use notes section for
additional information.

FIGURE 10.3 Documentation Estimating Worksheet

The Documentation

Major Chapters/Manuals		Miscellaneous

❑ Assembly   ❑ Preventive maintenance   ❑ Parts lists

❑ Installation   ❑ Corrective maintenance   ❑ Training syllabus

❑ Functional theory   ❑ Overhaul and repair   ❑ Recommended spares

Illustrations and Estimated Numbers

❑ Assembly drawings:_____   ❑ Exploded views: _____

❑ Schematics: _____   ❑ Plan views: _____

❑ Illustrated parts breakdowns: _____   ❑ Pictorials: _____

Estimates

Number of text pages for:

Assembly:_____   Preventative maintenance:_____

Installation:_____   Corrective maintenance: _____

Functional theory: _____   Overhaul and repair: _____

Training materials: _____   Miscellaneous: _____

Notes: _____

_____

_____

_____

_____

Instructions: Cross out items that do not apply. Use notes section for additional information.

the rare—applicants who mastered the test so aptly were those who saw words as only one of the tools of communication, and who therefore had the wit to first draw a little sketch of a check already made out and then write their explanation with reference to the sketch. The rest was easy!

That points up just one of the many problems with user instructions: Too many writers seem to have the notion that their craft is confined to the use of words alone so that many writers tend to use illustrations, if they use them at all, only as an afterthought, appending the illustrations to the text instead of basing the explanation on the drawing or photograph. In short, they illustrate their words instead of illustrating the concept. They forget that words are only symbols, are quite abstract as symbols and are not nearly as effective in communication of many images and ideas as graphic illustrations are. And in technical writing especially, graphics and other types of illustration are one of the most valuable tools, if not the most valuable.

Illustrations Are Not Always Graphics

Helpful as photographs, drawings and charts may be, they are not the only kinds of illustrations that can be resorted to and that prove helpful. Ideas also are often illustrated—even illuminated—by such writers' imagery as analogies, metaphors and similes. It is perfectly accurate and correct to explain that the earth is an oblate spheroid, but those words would send a great many readers to the dictionary and cause many others to sigh and turn to something else. But most would have no difficulty in visualizing the earth as "round and flattened slightly at each end, like an orange," because most of us are quite familiar with oranges. But if there is some good reason to use the more technical and more abstract term *oblate spheroid,* a simple line drawing illustrating what that term means would be equally satisfactory.

On the other hand, some concepts cannot be illustrated so easily, even by metaphors, similes or drawings, but that does not mean that something cannot be done to help the reader. Some analogies can be devised to help a reader understand an abstraction about an unfamiliar proposition.

Too many technical writers fail to make use of matrixes and tables to aid understanding and offer a more efficient presentation. In one case, a consultant converted an almost endlessly repetitive narrative describing a series of maintenance procedures into a matrix. That not only reduced the sheer bulk of verbiage to a fraction of the original draft, but simplified the problem of understanding the material, because common factors were revealed and principles were dramatized in the matrix format, which is inherently more efficient than narrative.

The original technical writer was astonished. He had, for some reason, assumed that only numbers and symbols were suitable for tabular organization, so it had never occurred to him that the narratives could be assembled into groups and blocks of text with common characteristics and organized into a table or matrix.

There May Be a Literacy Problem

One of our more stubborn problems is that of illiteracy. And literacy or illiteracy is not a black-and-white proposition, but exists in stages or degrees of literacy. The term *functionally illiterate* has been increasingly used to describe those who read haltingly and have extremely limited vocabularies. This has taken a curious twist in modern times and become something of a phenomenon. Even well-educated people today are often less than totally literate: they read poorly, despite high school diplomas and even college degrees. We also are still largely a nation of immigrants, and so many among us are not highly literate in English, but may be quite literate in their native languages. Thus we must always consider this factor, especially when writing instructions concerning ordinary consumer goods.

Technical Writing Does Not Have To Be Stultifying

I had the temerity once to begin a technical manual about a technologically advanced missile system—the tracking and guidance radar equipment in the system was of revolutionary design, with a device known as a Luneberg lens—with a writer's tactic calculated to gain the reader's attention and arouse interest immediately (known in the writing trade as a "hook"). I did this by presenting a summation of the system's unprecedented and near-miraculous capabilities and the promise that pages to follow would explain in detail how these marvels were accomplished by this breakthrough design in radar and missile control/tracking technology.

A design engineer assigned to conduct technical review of my manuscript condemned it with the observation, "It reads like a [expletive deleted] novel," which he obviously intended as a scathing denunciation.

This individual took it upon himself to pass judgment gratuitously on that for which he was unqualified—editorial merit—at the expense of technical review. It seems that everyone wants to be an editor and thinks himself or herself qualified for the task. And in their eagerness to impose their editorial judgment, those individuals often tend to neglect that review responsibility with which they are properly charged. Worse, however, is this unfortunately typical attitude of many professionals, especially the technical experts whose duties involve them somehow in the development of the user manuals. Their bias is that the writing must be wooden, the pace laborious and the exposition totally unimaginative, although they surely would not recognize these as fair characterizations because they confuse formal and dignified prose—which they apparently think is essential to their professional images—with such writing. They are unaware of or indifferent to the fact that some of the world's most eminent scientists and thinkers—e.g., Sir Arthur Eddington, Bertrand Russell and George Gamow—wrote serious, formal books in enormously readable style and with brilliant humor.

It is an attitude that results in many technical publications being difficult to read and even more difficult to understand without strenuous effort. It is also a completely unnecessary and irrational bias: user instructions do not have to be ponderous and difficult to read, and there is no earthly reason that they should be so.

Consequently, one of the objectives of this chapter is to persuade you to the philosophy that user manuals should be written in more popular style than most have been, and that will actually improve these manuals.

But It Is Not Entirely *How* You Write . . .

One serious obstacle in the way of good manuals is the universal tendency to refer to *writing* manuals as though the act of putting words on paper represents the entire task or even the major part of the task. Even when we recognize that writing includes illustrations of many kinds, graphics as well as imagery in language, we must recognize that *what* you write in a manual is even more important than *how* you write it. Even a well-written manual that neglects to include necessary information is not a good manual.

Recently, the calendar that is an integral part of my computer advised me that we were in the year 2005. I had no difficulty changing the date to the correct one for the remainder of that session, but when

I restarted the computer on a subsequent session I found myself again in the year 2005, as far as my computer was concerned.

I searched the manufacturer's manual and an expensive one I had bought in a local bookstore for a command that would correct the computer internally and return it to the present year. All I learned was that the time and date functions reposed in a file named PC TIMER 1.3; nothing more. I found the answer eventually, but not in any of the manuals where I should have been able to find it. I was compelled to resort to a highly technical method the average user is not likely to be able to employ. The information should have been included in the manufacturer's literature, of course.

What to write—a complete inventory of the user's information needs—can be developed only through careful research and planning, the amount of which is dependent primarily on the prior knowledge of the writer. It is not a simple process. It is, in fact, usually an iterative one, between the two interdependent main elements of research and planning. It begins with an assessment of the need—for whom is the manual intended and what is the user expected to be enabled to *do* with the information—in as much detail as possible, then the development of a preliminary outline, followed by as many iterations—more thought, more planning—i.e., expansion of the outline—more research and so on, as in Figure 11.2. And that latter function, research, is necessary even for the subject-matter expert, who conducts at least part of the research by exploring his or her own knowledge while expanding the outline. Each iteration of one function leads to an iteration of the other until some point where you believe you have described in your outline precisely what is to be included in the manual. Only then are you ready to begin "serious" writing of a first draft.

Summation

Among the many messages packed in this too-short opportunity to discuss this important subject of writing for the independent contractor has been one mentioned earlier and as often as possible: The need for planning and preparation. Good writing does not just happen; it is planned, and even then first drafts are rarely good enough. Good writing is as much the result of *rewriting* as it is the result of careful, thoughtful advance planning and preparation. A mistake all too common among those still new to the process is that of attempting to write a first draft too soon—before planning and research are complete or appear to be so.

(It is not at all unusual to discover the need for more information—more research—while writing, and that additional research may make the essential difference between an excellent manual and a marginal one.) Probably at least one-half of the time ought to be expended in this planning and research stage, except in unusual circumstances. It is time well invested. To a large degree this is the "secret" of writing, for it is the lack of planning and preparation that makes writing difficult. With adequate planning and preparation, not only is the manual a far better product; the writing of it is then a far easier task.

11

Hazards in Contracting and Methods for Combating Them

The chief trick, and the most difficult one, is that of getting it right the first time.

Independent Contracting as a Custom Service Business

All business ventures entail hazards, and certainly custom services businesses qualify quite well as businesses in which hazards lurk. Few businesses offer greater opportunities to lose money than does the custom service business. A great many kinds of contracting businesses involve the provision of relatively routine or even cut-and-dried services that do not vary greatly from one project to another. That is not true of all kinds of contracting, of course, because each project is unique to at least some degree, and so the element of unpredictability is always a potential. Therein lies the potential for disaster, the natural hazard of contracting, independent or otherwise.

Quite often, your new project will turn out to be something of an adventure into new, often uncharted territory. Some projects are of that description by their nature, others because of little reliable information up front. That is, the contractor's bid or proposal is often based on the client's description of need, but the client may or may not have had a clear and accurate idea of his or her need. If you, in that case, you are basing your plans and your commitment—your bid or proposal, that is—on that sketchy and questionable information, you are basing your estimate on only partly known factors, and often with your fingers crossed. In trying to beat the competition and win the contract, while hoping for the best, you are gambling. (It's common enough, but that does not make it less hazardous.) You therefore should plan escape routes in case your effort to be competitive proves to have been based on an estimate that was unrealistically low. Or perhaps there are other approaches to protecting yourself against disaster.

It has been my fortune to be employed by companies where that has been the case. One firm bid and won custom service projects that were doomed from the start. In postmortems on such projects, there was never a dearth of excuses and alibis, the most frequently heard being that the original estimate was too low. That is uttered by those who had nothing to do with the estimating, of course. But my own experience convinced me that low estimates have much more often been the excuse for, rather than the true cause of losses in custom projects. Much more often the culprit is underperformance, rather than underestimating. All too often, a manager hungry for a contract will gamble on a low estimate, with little planning or unclear idea, hoping somehow to muddle through. The problem is a grand deficiency in preparing for, doing and managing the work, in a program including the three following general ingredients:

1. Thorough, detailed analysis and planning before making the estimates. (That alone greatly reduces the hazard of underestimating.)
2. Complete documentation of the planning and specifications.
3. Close control of the project, to include:
   - A detailed monitoring system, with clearly identified milestones.
   - An early warning system, to sound the alarm as soon as the achievement of an objective is threatened.
   - Identification of all possible contingencies.
   - Recovery plans to cope with all contingencies.

What Is a Custom Service?

Although contracting is inherently a custom service, I will use the term here in a special sense: It denotes a service that is nonroutine, a service in which you, as the contractor, provide your services in a project that is unique in some highly significant way that requires you to devise a specific plan for the undertaking. I therefore classify contracted services as either custom services or general services. The following examples will clarify this.

A customer may engage the services of an established polling organization, such as the Roper or the Harris organizations, for a study along the lines of its usual work, employing its established resources. In a strict sense, this is custom work, since it is being done for a specific customer and to satisfy a specific need. Since it is a routine service performed by an organization established especially for that service, however, it usually entails few uncertainties and no special design efforts. Therefore, I count this as a general service.

On the other hand, should the customer require a special poll or survey, for which it can supply few specifications other than identifying the kinds of information required or the problem being addressed, the bidder ordinarily has to do special studies, probably prepare some kind of proposal and chart a path through new territory. I characterize this as a custom service.

The distinguishing characteristics, then, are that the custom service typically requires R&D (research and development) or at least special design efforts, usually some pioneering effort and some substantial degree of uncertainty as to problems to be overcome and costs to be anticipated. These kinds of projects typically entail R&D programs, research, studies, design and prototyping equipment, special consultants, progress reports and other such efforts and functions. As a result, they are usually high-risk projects. Even with cost-reimbursement contractual arrangements, there is some degree of risk. The customer is not certain what the final costs and outcome will be, and the contractor is not certain that he can overcome all obstacles and realize a profit or, for that matter, avoid loss.

This is why so much advance planning and so many plans for close management control are necessary to handle such projects at a profit. Profit aside, however, for the small, independent contractor, the main concern must be survival, getting the job done successfully and without operating losses.

This is why so much advance planning and so many provisions for close management control are necessary to handle such projects successfully. You must anticipate costs and possible cost increases, if the project is to be a lengthy one. And you must not make rough guesses in estimating costs: Today, "sticker shock" is all too common for all too many items.

Characteristics of Custom Services

In most cases, custom projects are labor intensive, although in some cases there may be a fair amount of cost resulting from the use of equipment and material. In most cases today, costly labor is required, usually heavy with technical and professional specialists, frequently supported by other professionals whose labor is not inexpensive either. So these projects are often labor intensive in both senses: Labor is the chief item of cost, and it is labor at high hourly rates. And since these are typically high-risk projects, you must always ask yourself about the exposure or downside before committing yourself to your final cost estimates.

Particularly characteristic of custom service projects is that most call for creation of a unique project plan, developed specifically for the project. This in many ways is the Achilles' heel of these projects. Typically, the project has been conceived, and perhaps presented to the customer in a proposal, with idealized projections of events and their dates. Once a proposed project has become a contract, reality sets in. You must now deal with the realities of the budget and resources available, the latter often far less than ideal.

There is at least one other factor of great concern. In conceptualizing and proposing the project, you have used your imagination to paint a picture that is not easy to bring to life—e.g., you promised to be supported by an outstanding consultant of some sort, but the budget really can't afford that consultant. Idealizing the proposed project to maximize its sales appeal may itself be a trap if you are not careful to apply the brakes. Do not allow your enthusiasm to carry you away.

Finally, an inherent problem in many kinds of custom services is that it is difficult to draw precise specifications of what is to be done and, especially, of what the final product is to be. In such projects, much depends on the client's subjective reaction to the result you present. For example, if you are retained to design and produce a company's newsletter, you are almost entirely dependent on the client's judgment. To minimize the potential problem of client rejection of your efforts, an important characteristic of what I call a custom service here is that there is at least one in-process review, and perhaps two or three.

The more reviews you have during the development, ordinarily, the less difficult it is to win the client's approval of the final product, if you have responded well to the client's reactions at every review. It is thus wise to encourage as many reviews as possible. Even an unreasonable client will find it difficult to object to a product that is as he or she specified a wish for during intermediate reviews.

In this connection, do not explain your problems or hardships during client reviews. It was once pointed out to me, as a young and untried manager, that "Nobody"—meaning the boss—"cares about the storms you met at sea; they want to know only whether you brought the ship safely to port." It is a poetic way of saying, "I don't want to hear about your problems or your great efforts; I just want to see results." (Substitute "client" for "the boss" in the above.)

When I first heard this, it seemed a cruel and cynical observation. Only later did I recognize that it was a great kindness to advise me of a harsh fact of life, which Harry S. Truman expressed by his own homily: "If you can't take the heat, stay out of the kitchen."

An independent contractor is judged as a manager is: by results. That is fair enough, I think, since it is results that the client wants to buy. It doesn't matter to the client how easy or how difficult it was for you to get those results.

Precontract Planning and Estimating

The question naturally arises of how much planning can be done before the contract is signed. How much is economically feasible, and how much is absolutely necessary to reduce the risk to an acceptable level? The answer lies in how much information you need to make an estimate with reasonable confidence. That defines how much planning must be done in advance of the actual contract. More than that is wasteful and increases overhead costs unnecessarily; anything less than that introduces unacceptable risk.

The more or less traditional injunction against reinventing the wheel is usually meant to be humorous. Again and again, however, technical and professional people are guilty of doing this. Perhaps this is due to simple ignorance—they do not know what has been invented. Perhaps they think that their own design will be far superior to anything extant. Whichever is the case, the irony is unaffected. Reinventing the wheel is a technological and economic sin that is committed all too frequently, and it can be responsible for an otherwise well-conceived project becoming a disaster.

Once in writing a proposal for a vocational training center, I was required to include detailed outlines for training programs in six vocational trades. I searched out the most well-written how-to texts in those trades I could find and converted each text into a detailed outline. I bought an encyclopedia from the Government Printing Office of audiovisual programs available from various sources and selected the most appropriate of these to support our program. Using these and similar tactics, I produced a three-volume proposal (about 1,000 pages total) for about $12,000. It would have cost about four times that amount without the use of the many "wheels" that had already been "invented" for my use. Always investigate what wheels have been invented and are available for your use before you propose to invent your own.

Planning and Estimating Methods

Some time ago I decided that one of the best ways to plan and estimate a project is to prepare some sort of functional flowchart for it. A graphic presentation helps you visualize the project in all its workings and helps you identify the significant phases and functions. It compels you to think the whole project out, in fact, rather than fall back on hunches and intuition.

For some reason, I am prejudiced against those top-to-bottom flowcharts of which computer programmers appear to be so fond, possibly because as a child I was taught to read from left to right and I instinctively follow that convention in charting as well as in reading and writing. So all references to "right" and "left" made in connection with charting mean that "right" is further downstream, toward the finish line.

It is possible to start such a sketch at either the left- or right-hand side—that is, at either the start (first steps) or finish (final steps) of a project—depending on certain factors. Usually, in a custom project, it is possible to sketch in both ends immediately because you generally know the last step, the result or product you are to produce; and the first step, what the customer has specified and what you must do to initiate the project. If the client has given you a set of performance specifications, your design must translate those into manufacturing or physical specifications, as shown in Figure 11.1.

As an experienced practitioner of your profession, you know the typical first steps of such a project, and you know the normal practices and procedures typical of such projects and the normal sequence of steps (unless the project is an atypical one or is atypical of the projects you would normally undertake). Therefore, it is fairly easy to sketch in the main steps quickly, as exemplified by the hypothetical project in Figure 11.2.

FIGURE 11.1 First Steps of a Project Shown in Flowchart Form

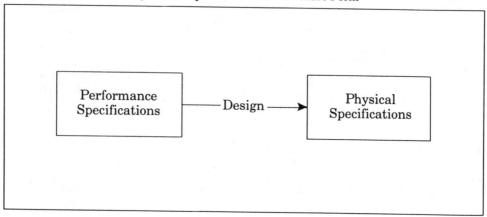

Consider that moving from left to right along the line and in the direction indicated by the arrows represents time, as well as a logical sequence of functions. You see that the flow already begins to represent a plan of action, although only the major functions are portrayed here. You can easily develop this chart further, into a more detailed chart, by breaking each of the boxes into the flow of actions necessary to perform the major function represented. You would also build feedback loops into the process, as shown, since the writing of a first draft would usually reveal the need to return and further refine the planning procedures of earlier steps. This makes it an iterative process, characteristic of true automation, wherein output information is fed back to input as a correction signal.

To translate the client-supplied specifications of performance and perhaps even some physical characteristics, you can begin by translating them into a flow chart of this type, beginning with the major objectives and adding detail until you have a sequential flow that is a design of the process. The use of this graphic method takes most of the guesswork out of converting the client's words into a working plan. Once you get those words converted into a flowchart, you can begin an analysis and develop a working plan to serve you both as the basis for a proposal and a project plan budgeted properly. One of the tools available for this is a discipline known as *value management*.

FIGURE 11.2 The Iterative/Reiterative Generation Process

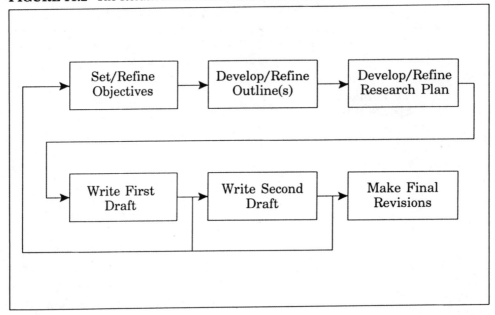

Value Management

The concept of value engineering was born shortly after World War II, as a result of some wartime experiences at the General Electric Company. An executive in that company discovered that in many cases, when the company had been forced to find and use substitutes for scarce materials, the substitutes had proved to be superior to the originals. He began to find more and more instances in which this was true, enough such instances to suggest that this was worth investigating further, when time and circumstances permitted.

After the war ended, he assigned a young GE engineer, Larry Miles, to investigate the phenomenon. Miles verified that almost everything manufactured could be improved by substituting material and eliminating unnecessary parts, changing processes and other actions that reduced costs without reducing the quality or the utility of the product. A part that was made by machining down a block of steel might be manufactured just as effectively by stamping or molding, at far lower cost. One of the new plastics might perform as well as or better than the original metal. Components might be eliminated without affecting the basic utility of the product.

Such changes meant that value was added to the product. The product was as useful as ever and of as good a quality, but simpler and less costly to manufacture. So, with the philosophy that reengineering the product added value to it, Miles called his developing discipline "value engineering"—engineering to increase value. He had developed an organized system, a specific methodology for engineering greater value into products and he was ready to teach others how to improve value through deliberate effort, rather than though lucky accident.

In later years the system came to be known as "value analysis" and, more recently, as "value management." The latter term was adopted by the General Services Administration of the federal government to deemphasize the engineering aspects, since the methodology is applicable to a great many things—to almost any endeavor or function, in fact. The use of the earlier term had tended to frighten off and intimidate those with no technical training; therefore, the term *value management*, or VM came into use.

There is not much point in going into all the implications of VM, but the basic principles are of great value to everyone, and especially to many of the functions and problems that are part of independent contracting. But first a few general observations

Value engineering or value management was designed originally to be put to work by a group of people working as a team, and it is still formally used in just that manner—by assembling a specific team to analyze a product and devise improvements. In principle, however, anybody can apply the methods of VM. In fact, one handicap VM proponents have in persuading others to adopt the system is the stubborn belief by many people that they have always employed its principles—finding the most efficient methods and materials—in their work. Many resist the idea that VM is anything new or different.

The system is entirely logical, and depends heavily on positive thinking: true reasoning, in the most literal sense of word. But recognizing that most of us, fallible human beings that we are, have great difficulty in being completely objective and rational, the system employs certain methods to ensure objectivity, and they will ensure objectivity if you adhere to them scrupulously.

To date, VM has been the technique most commonly employed to improve existing products and systems, whereas it ought logically to become even more useful as a guidance to the design and development of new products and systems, to original design, rather than redesign. VM should help you get it right the first time.

Function

For some reason, nonengineering and nontechnical people in general appear to be awed and mystified when they learn that VM depends heavily on the basic concept of identifying or defining functions. In common usage, *function* can mean a number of things: a formal banquet, a mathematical constant, an official position or a purpose. In VM, there is no great mystery about the meaning of the word. It is what a product or system is supposed to do: the reason for the existence of the item, where *item* means product, component, system or even job, such as secretary.

The function ought to be defined, according to Larry Miles' nowaccepted system, in unequivocal, objective terms. To ensure this, the ground rule is that a function can be defined in only two words, a verb and a noun. In a few cases, where absolutely necessary, more than two words may be employed, but they must be verbs or nouns, not adverbs or adjectives. That's because verbs and adjectives tend to introduce emotional, subjective ideas, qualifying and compromising the bare and basic definitions sought.

As Gertrude Stein put it, "A rose is a rose is a rose." No adjective can change that fact, and modifying it by such a word as "beautiful" or "breathtaking" is simply supplying an opinion, not an additional fact. Of course, it might be compounded as a red rose, if there is some need to specify that it is red. But then it ought to be a compound noun—red-rose—rather than a modified one (philosophically, at least, regardless of grammatical rules). Let's look at a few examples:

Item	*Function*
paper clip	binds papers
ashtray	holds ashes
adding machine	adds numbers
calendar	presents dates
correction fluid	covers mistakes
diploma	documents graduation
lock	secures door

Simple, isn't it? But what's the purpose of this simple exercise? Let's look at some more complicated examples. For example, what is the function of an ordinary pencil? It "writes words," true, but that is not an acceptable functional definition, because it may also be used for drawing, making check marks and many other things. Therefore, it is far more basic to define the function as "makes marks."

How about that little check-writing machine that many offices have? What is its function? A quick answer would be "writes checks." But once you have begun to think like a value manager, that won't satisfy you because it's not the *basic* function of the machine—not the *purpose* of having the machine. You can write checks in longhand or by typewriter as quickly and as legibly, and spare yourself the $150 or whatever the machine costs to own.

Why does one buy and use such a machine? Is it not because the machine makes perforations where the amount of the check is inscribed, making it difficult to feloniously raise the amount of the check? In short, the function of the machine is "protect checks."

The moral here is a simple one: When trying to identify or define a basic function, think in terms of purpose. Why does the item exist at all?

Basic versus Secondary Functions

We have been talking about defining the basic functions of various kinds of items, but in many cases items serve more than one function. It would be possible, for example, to design a check protector that would do nothing but perforate the area of a check in which the face value was written, after the check had been made out by hand or typewriter. Such a check protector might be cheaper to manufacture than are current check-writing machines, but it would also be less convenient to use. Therefore, check-writing machines inscribe the amount in ink, as well as perforating that area so as to make it all but impossible to change the amount inscribed. That is, the primary purpose and basic function of the machine is to protect the maker of checks, but it has at least one other function; it "writes checks" too. Bear in mind, however, that the latter is a secondary or supplemental function.

Some items have many functions. A watch "indicates time," but it may also serve as a calendar to "indicate dates," have an alarm to wake you up and have some built-in extras, such as a thermometer and a barometer, as my office clock does. "Gives temperature" and "predicts weather" are additional or secondary functions of my office clock.

In many VM efforts, secondary functions are eliminated because they are found to contribute little or nothing to need, while they add cost. It is important to identify all secondary functions, as well as the basic one, in order to weigh their contribution and make value determinations.

The basic VM system, then, is largely one of analysis, employing a series of questions and structuring answers according to preestablished rules. There are three phases to VM:

1. This phase is called the analysis phase, for obvious reasons. The prime purpose of the first phase is to identify and define functions, basic and secondary.
2. In the synthesis phase, the results of the first phase—the objective, crystal-clear picture of functions—are used as the basis for design or redesign of the item, to increase value or reduce costs without reducing quality, reliability or required utility.
3. In the implementation phase, the results of the earlier phases are put to work to make the design or redesign a physical reality.

The Right Questions

Obviously, before one can get answers, one has to ask questions, and only the right questions bring forth the right answers. You've learned the first question. What is the basic function, or purpose, of this item? And you've learned that it is usually wise to ask yourself what the need is to understand your own questions about purpose and function.

Formally, the questions proceed somewhat as follows:

- What is it? (Name)
- What does it do? (Basic function)
- What does it cost? (To fulfill its basic function)
- What else would do it?
- What would that cost?

You should also ask, "What else does it do?" referring to secondary functions, what they cost, how necessary they are and whether they could be accomplished by some less expensive means.

Once you have mastered the art, you will find that VM can be used to economize on any parameter; that is, it can be used to economize on time or materials rather than money alone. For example, when the Environmental Protection Agency found that the processing of grant applications for wastewater treatment plants was taking so long that the money would not be spent within the time period mandated by the law, it instituted a VM program to determine how the process could be speeded up. I participated in the VM study that revealed the problems in the system responsible for the delays. In other applications, the VM system is used to find energy conservation expedients and reduce reliance on scarce materials. Instead of "What does it cost?" you can ask, "How long does it take?" "What else will do it?" and "How long would that take?" when the aim is a shorter timetable.

Apply this method rigidly to analyzing client's problems, designing and proposing projects and carrying out your projects. Ask the questions constantly, not only when designing a program, but as you go, always seeking a better way: Do you really need all those forms, or can one be simplified to do the work now done by three or four forms? Are you keeping any records because "it's always been done this way," or can you eliminate some without loss to anyone? (It wasn't until after World War II that someone in the British Parliament decided to find out what the purpose was of that artilleryman who, by tradition and long practice, stood to one side every time an artillery piece was fired, and who appeared to have no other duties. It was discovered that, in times long ago, the job had been to hold the horses so that they wouldn't bolt when the gun was fired!)

Defining Problems

It has become something of a humorous platitude in some business circles that there are service firms who offer solutions for which there are no known problems. Like many humorous platitudes, this one is rooted in truth. There are indeed those who insist on defining every problem in terms of the solutions they have already devised. Samuel Neaman was remarkably successful in turning the McCrory Corporation around from years of declining business and losses to an extended procession of growth profit. But when he attempted to use the same methods to revitalize the S. Klein department store chain, he found them inadequate to the task. S. Klein was not a chain of variety stores, and successful merchandising and management methods for a variety store simply did not do the trick. Obviously, the problems of S. Klein were quite different from those of McCrory and should have been attacked with quite different approaches.

One factor that often stands in the way of our getting a proper definition of a problem is our own inability to be objective and abandon our biases. Henry Ford went from the four-cylinder Model A Ford to the eight-cylinder V-8 Ford, insisting that a six-cylinder engine was scientifically unsound, even after his competitors had marketed successful six-cylinder engines. Virtually no advance—not the electric light, not the automobile self-starter, not the skyscraper, not the high-speed turnpike and not even the computer—has been achieved without sincere critics who see insurmountable problems preventing its realization or insist that the basic idea is unsound.

To even define a problem, must less solve it, one must first somehow lay aside prejudice and approach the problem with an open mind, as did an anonymous physician who first said, "There are no incurable diseases; there are merely some diseases for which we have not yet found the cure." The mind-set must be that there are solutions to all problems, and they can be found.

What you need first is information: the facts that define or identify the problem. Here is a logical problem-defining and problem-solving process:

1. Identify or define need or goal. (For example, reduce the time or effort required to record employees' time.)
2. Review and analyze current system. Make a functional analysis. Why do we need time records? What are they used for? In what form must records be? What happens to them next?
3. Identify basic function and subordinate functions.
4. Evaluate need for all subordinate functions. Can any be eliminated without compromise to basic function?
5. List all possible alternate ways of performing the basic function.
6. Evaluate all alternatives listed.
7. Make final judgments.

Let's use a simple example to illustrate this process in action, based on an actual case in a company where we were still using 80-column IBM punch cards to input data into our mainframe computer.

Your company assigns an item identification code (job number) to every project and to other cost centers, such as sick leave, vacation, overhead work and idle time. Each employee is required to record how he or she spends the eight-hour day by writing the job numbers on a sheet, with notations of how much time was spent on each. Each employee has a simple form, issued daily and gathered up by the direct supervisor at the end of each day.

The accounting department prepares a master sheet each week, on which all these charges are accumulated. These weekly summations become the basis for ledger entries and various project reports. Each person's time record also becomes the basis for the employee's master file, for payroll and related purposes.

All of this is handled by computer, with the data from the various time records entered on punch cards and the cards entered into the computer, which keeps track of all charges and issues pay checks every week. You decide that the basic function is to furnish data for payroll and various reports. The data must be in some form that is "machinable"—able to be read by the computer. That's accomplished by punching the data onto 80-column IBM cards. Investigating, you find that the data are entered into the computer weekly. Why, then, should the employee make out a new card every day and require the accounting people to compile it into weekly totals for punching? Would it not be simpler to have each employee keep a weekly sheet, and turn it in at the end of the week, ready for the supervisor's approval and transmittal to the keypunch operators?

Say the supervisors want to keep close track of charges and costs for every cost center and project. They want the figures every day. They claim a week is too long to wait to see accumulating charges, especially those for nonbillable hours. How can the employee keep a weekly time record, yet give the supervisor a daily account of the day's expenditure of time?

Two possible answers to that occur immediately:

1. Duplicate the time sheet daily and give the supervisor the copy. But that means both the extra cost of duplicating and the extra cost of time spent, with everyone trying to gain access to the copiers at the end of the day.
2. Use carbon paper and give the supervisor a carbon copy of the day's entries. The problem has been identified, and at least two possible solutions synthesized.

In the actual case, another problem cropped up. Many errors were made in keypunching data from the sheets, especially because the sheets were clumsy to handle and employees sometimes failed to identify clearly whose time record was being presented. This was solved readily, and another efficiency introduced, by sending each employee an individual punch card, already inscribed clearly with the employee identification, and having the employee write time charges directly on the card. The keypunch operators then punched the card itself, and the card reader input device read the punches, ignoring the pen or pencil marks, of course.

The key to problem identification is analysis, and analysis is a fact-gathering process. That same Sam Neaman who pulled McCrory up by its own bootstraps railed against making assumptions; he even had posters made up exhorting McCrory employees to "burn" the word *assume* from the McCrory lexicon. Instead, he declared, one needed facts. He insisted that the key to successful management included four essential ingredients: facts, plans, execution and supervision. We need analysis to gather the facts, plans based on those facts, faithful execution of the plans and intelligent supervision to ensure the execution of the plan.

Summation

The thrust in this chapter has been clear and objective reasoning, based on proved disciplines, to minimize the many hazards inherent in contracting, especially for custom services involving unique projects. However, the greater benefit of using these methods is that they add great power to your marketing campaigns. They enable you to plan better solutions to clients' problems, as well as better designs and more efficient programs, in general. In many cases, I have found, the client has a shallow understanding of his or her problem and is calling for the wrong solution. Your ability to make a presentation, written or verbal, demonstrating to the client not only your superior understanding of the problem, but your more efficient and probably less costly program makes you far more competitive than low prices and impressive brochures do. That is the greater value in doing a proper job of problem definition and program design.

Ethical Codes and Client Relationships

*There are written and unwritten codes of conduct for social situations,
and there are written and unwritten codes of conduct for business.
There are practical reasons to embrace these codes, as well as a
moral obligation to do so.*

Ethical Codes versus Client Relationships?

Ethical codes and client relationships are inextricably combined. In fact, ethical codes have more to do with client relationships than with any other element of running your business; they affect every aspect of the business relationship, including fee setting, marketing, contracting and negotiation.

In this era of the department store and the supermarket, whether for groceries or toys, little is left of the personal seller-buyer relationship that was once characteristic of trade. Today, we may become friendly with a checkout clerk or even get acquainted with a store manager, but we rarely meet a proprietor or anyone who is not an employee of the emporium. Employees do not make policy, and even managers and other midlevel executives are severely limited in what they can do to accommodate you when and if you have problems in your efforts to do business with the organization. The ethical code of the business is therefore whatever general rules and regulations have been established by management and what is required by law. It is a much changed environment and a much different set of conditions from that of only a generation or two ago, when the warmth and friendly understanding of the local merchant was the customer's assurance that the management—i.e., the proprietor—was concerned with doing the right thing and would do so, even when a customer was unreasonable, as some were.

With relatively few exceptions, that is a relationship of an era long past. What has replaced it is the sense that the large, well-appointed store or other corporation, with its aura of efficient management and organization, has no need to take advantage of the individual customer, but can afford to be honest and do the right thing. The aura of the big store and big company, supplies the only sense of security to the customer today. But that works only for the big store and big company, if it works at all. You and I must rely on the older, time-honored methods for winning the customer's confidence and trust: We must utilize the direct, personal contact, a code of ethics, diplomacy and good sense in all our relationships with our clients.

Caveat Emptor

It was not too long ago, less than 100 years, in fact, that "let the buyer beware"—*caveat emptor*—was the common and universally accepted code of the business world. All sales were final, and that was that. Young Montgomery Ward shocked and outraged the older and more experienced businesspeople of his day when he introduced the unconditional guarantee in his fledgling mail-order business. His frightened competitors and contemporaries immediately denounced him as a fool who soon would be bankrupt as a result of his insane notion of guaranteeing his customers satisfaction or their money back with no questions asked.

Implied Warranties

Of course, the unconditional, money-back guarantee is such a standard provision of business today that most customers assume without asking that they can return anything they buy that proves to be a disappointment in some way. In fact, the laws support them by inferring that in common law there is an implied warranty of any goods purchased in good faith. Thus an implicit and tacit code exists, with respect to warranty, under every state's Uniform Commercial Code. Customers are entitled to assume that the goods they purchase are suitable for the use intended. Store owners know that their customers assume they have this right, and many stores have notices posted on their premises to explain the terms of their guarantees. These usually specify some time frame within which the goods must be returned, and often provide that only store credit will be offered, but no cash refunds will be made.

This general, implied warranty applies specifically to the sale of goods and is quite separate and distinct from the express warranty that an individual dealer or manufacturer might offer clients. It does not apply to the services themselves. Contract law normally covers the matter of a guarantee of satisfactory delivery or performance of services contracted for. However, the question of warranty or guarantee is only a small part of the ethical code I would urge on you as an independent business owner and contractor. The ethical codes to which you subscribe will be a key element in the reputation you build. A good reputation—i.e., your good name, as it has been put so often in the past—is as important an asset to you as a poor reputation is a serious handicap.

Caveat Venditor

Actually, legislators framing the modern laws governing business today have shown great awareness of the many ways that the more predatory and clever business owners among us have taken advantage of the less experienced and more trusting customers. They have, therefore, gradually developed a comprehensive structure of many laws to protect buyers from their own innocence, as well as from the unscrupulous business practices that once were commonplace. These laws have grown to such a degree that many individuals in trade and law practices have observed that we have almost a *caveat venditor*—let the seller beware—situation today. It is distressingly easy today for even a well-intentioned and cautious business owner to run into legal problems unexpectedly.

Even that isn't enough. Despite that protection under the law, it is important for every business owner to subscribe to and observe some principles that constitute a code of behavior characterizing the individual and the business: an ethical code. It is especially important for you, as an independent contractor, to do so. For one thing, as an independent contractor, you depend or should depend on word of mouth—referrals and recommendations—to find new clients and contracts. Word of a good reputation travels fast, but not nearly as fast as that of a bad reputation. Thus, becoming known as an ethical, conscientious and dedicated contractor is a great asset in winning new contracts, but it is an even greater asset in overcoming and offsetting the almost instinctive skepticism of the buyer in today's world. Bear in mind that today's consumer, unless very young or very naive, tends to a natural caution, especially in contracting for services. Perhaps the customer once started with the premise that the seller was honest and honorable, but you cannot assume that attitude on the part of a client today. Today, you must work at winning the client's trust and confidence.

Contracting often places you in a special position with regard to the client and his or her affairs. In many kinds of contracting, such as training a client's staff in the use of some management system you have written for the client or perhaps installing a marketing program for the client, you become privy to confidential detail and proprietary information regarding your client's business affairs. That places a special obligation on you. You must recognize that you do have access to highly sensitive information about your client and your client's business, and you must be immediately alert for any conflict of interest.

Conflicts of Interest

Conflicts of interest are among the most common issues that any ethical code of conduct must address. Conflict of interest is easy to define: It is a situation in which your personal interest is at odds with or opposed to your client's interest, although it may be very difficult to cope with.

The temptation to use confidential or proprietary information for your own gain is a strong one, and often difficult to resist. Several situations can easily create a conflict of interest for you, if you permit such conflicts to arise. Most formal contracts specifically enjoin one or both parties to respect and preserve the other's confidential and proprietary information. Here are a few typical situations to illustrate the problems that may arise.

Building on a Competitor's Strategy

You are providing Mr. Smith copywriting services to support a sales campaign in a highly competitive industry. You have recently worked for Smith's direct competitor, Ms. Jones, writing promotional sales copy for a directly competitive product. As a result, you know what Jones' sales strategy is. Can you use this information ethically to make Smith's copy more effective?

This is a very difficult situation. There is no way you can unlearn what you have learned recently, but neither is it fair to Smith to deliberately deny him the best copy you can write. Probably you should not have accepted two clients with directly competitive businesses, since you work in sales and marketing, which almost guarantees that you will encounter this kind of conflict unless you are quite careful. You do have the problem, however, and the only thing you can do here is to produce an entirely different strategy for Smith. You must somehow manage to do a good job for Smith without being influenced by what you know about his competitor.

Reveal Proprietary Information—or Else

You are an accountant providing audit services for Mr. Smith. Smith knows you have done similar work for Ms. Jones. He asks for information about her costs, mark-ups and other fiscal data. He swears that anything you reveal will be held in confidence, and no one else will ever hear of it. When you resist revealing what you know about Jones' business affairs, Smith threatens to cancel your contract immediately.

As strong as the temptation might be to submit to Smith's threat in order to keep a profitable contract, you must resist it. It would be entirely dishonest to leak such information, and that ought to be reason enough to refuse his demand. But it could also be suicidal. How much trust would Smith put in you if you told him what you knew about Jones' affairs? (How do you know, moreover, that Smith is not testing you to see if you can be trusted with knowledge of his business?)

Working with Competitors

You are under annual retainer as a proposal writer to Smith & Company when Ms. Jones calls and asks you to help with a proposal for Jones & Company, a direct competitor. Can you accept the assignment?

The proper thing here is to call Smith and ask if he is going to pursue the contract that Jones is pursuing. If Smith advises you that his company will not pursue that contract, you are then free to work for Jones, but only in that case.

In all these situations and similar ones, where you gain proprietary information about a client or even a prospective client—e.g., someone you chatted with at some business meeting when the other party was perhaps more frank than he or she should have been—you have a potential conflict of interest. First of all, you should try to anticipate the possibility of such conflicts and do whatever is necessary to avoid them. When, despite your caution, you get into such a situation, integrity is the only way out. Of course, you cannot be expected to avoid steadily increasing your general competence and the quality of the services you offer clients as the result of your growing experience. There is a not-so-fine line, however, between that and using specific data about clients for your personal benefit, direct or indirect, and you must not permit yourself to even stray across that line. In short, you must have absolute integrity.

Integrity

To have integrity is to be scrupulously honest and totally dependable, but it goes even beyond that. It is to do the right thing, which may call for more than mere conformance with what the law and even morality requires. There is too great a tendency today to equate that which is not illegal or immoral with that which is justifiable and acceptable, as though the statutes and common law themselves represented an ethical code and thus are adequate standards of behavior. Of themselves, the laws do not constitute a moral code. More than one scoundrel operates within the letter, if not the spirit, of the law, and still does harm to others. There is much more to ethics and to integrity than merely staying inside the strict bounds of the law.

Keeping Your Promises

Being scrupulous in keeping your promises to the letter is one of the most basic underpinnings of integrity in any professional or business relationship, even if it requires going the extra mile, as many situations do. Whatever I do for a client, I guarantee that I will do what I have promised to do, even if we do not have our agreement in writing. If I am working by the day, I usually agree to a not-to-exceed number of days to be charged to the client, while guaranteeing that I will finish the job, even if it requires more of my time than I had estimated. (Quite often, in proposal work, that requires a 24-hour workday in the final stages of meeting a difficult deadline.) If it proves necessary to exceed the number of days because I have underestimated the work required or overestimated how much I can do in a day, I will put in the extra days at my expense. Only if an overrun of days necessary to do the job is clearly due to something beyond my control, perhaps something the client has done or failed to do, or some contingency that could not have been anticipated, will I ask for a contract amendment and additional funding. If the cause of the need for extra days is unclear and my request is contested by the client, I will furnish the extra days at my expense and avoid dispute: I believe that my honor and honesty must be beyond reproach or even the slightest suspicion, and I will not put them in jeopardy for the sake of a few extra dollars.

That latter situation and my method of reacting to it represents leaning over backward to protect my reputation and is doing much more, perhaps, than is absolutely necessary. Still, I think one's reputation for fair dealing must be preserved, even if it means permitting yourself to be used badly by an occasional client whose ethics leave something to be desired.

Being Fair

A large element in creating a sense and an image of integrity is creating a sense and image of being fair in all your dealings. But what is fair? That is not easy to determine in most circumstances. As a seller, it is usually, but not always, in your best interest to decide the question of fairness in favor of your client.

In some cases, a client will push you to the brink in demanding as much as you are willing to concede. That client is often one who will have a lowered opinion of you if you permit him or her to continue to take advantage of you. Some clients respect and admire you only if and when you stop them in their tracks and insist on a reasonable degree of equity in your relationships. Be pliable and accommodating, but set a limit to how far you will go in permitting yourself to be used and abused. At that point, refuse further concessions, not adamantly or defiantly, but pleasantly, quietly and matter of factly. That approach demands respect and usually earns it without further difficulties. I don't know of a single case where I suffered as a result of being businesslike, as long as I was obviously being fair to all concerned.

One day I received a call from a client. More specifically, I received a call from a manager on the staff of a very large corporation that was our client. He asked me for additional copies of a manual we had written to accompany a training program we had developed for his company.

"How many more copies do you need, Pat?" I asked. (I knew the manager reasonably well.)

"About 300," Pat said.

"I will have to charge you for those," I said.

"Don't we have some copies coming?" he asked.

"Yes," I said. "You had 25 copies coming to you by contract, but I sent you 50. I could give you a few more copies, but if you need as many as 300, I must charge you for them."

Pat grumbled something about his need to look after his company's interests, I agreed that he should protect his employer's interests, but should also allow me to attend to my own company's interests. He then agreed to accept a bill for the extra copies of the manual. Pat agreed that my position was a fair one. No one's feelings were injured and we continued to do business together.

That exchange was not a dispute; it was a negotiation between contractor and client. Use the win-win method of negotiating anything that needs to be negotiated. That is a negotiating method that never seeks to gain an advantage of the other party, but always to find a formula for solving the problem to the satisfaction of both parties. The only good agreement is one in which both parties are truly satisfied. It is a bad agreement if one party feels used, abused or bludgeoned into submission.

Image

Integrity and fairness, as well as competence and dedication, are parts of the image you must project in consulting and contracting.

Professionals in the advertising world speak respectfully of *positioning*. By that term, they refer to the concept of yourself you have managed to create in the customer's mind—the image the customer has of you. One objective of advertising is to persuade the customer to see you as whatever you want the customer to see. In one advertising campaign, a department store may want customers to see the store primarily as a source of the best class of cosmetics, but another of its advertising initiatives may present it as a fine clothier, or as a vendor of furniture of excellence. On the other hand, store managers may choose to position their store as a one-stop source of everything. The position should be a conscious and deliberate choice, not an accident.

When I am selling my books and reports on writing, I want to be viewed as an authority on that subject, but when I am writing, lecturing and consulting on proposal writing, I must come across as an expert on selling to the government. If you do or sell more than one line of services or goods, you may need to have more than one position; as an independent contractor, however, the position must always be that of an ethical and honest contractor, as well as a competent one.

Dress Code

Not long ago, I found myself embroiled in noisy dispute with a bunch of computer software development contractors about how to dress on the job. Many of them defended quite vociferously the idea of jeans, tee shirts and sneakers for working on a client's premises, and even those who were obviously not in total agreement with that idea were reluctant to challenge it.

To a large degree, the orientation depended on geography. The California people (especially those from Silicon Valley and, to some extent, the Florida people) tended to the most informal garb for work, even in the offices of major corporations. They implied rather strongly that as sophisticated technical professionals, they were entitled to set their own dress code. It also proved to be a matter of age. The younger people were more likely to insist that jeans, tee shirt and sneakers costume was suitable office garb. It also somewhat depended on the profession, with computer specialists being most dedicated to that most informal dress code.

Perhaps I am irretrievably old-fashioned, but I side with the conservatives. I wear a jacket, tie, crisp white shirt, slacks and old-fashioned leather shoes to a client's office, and would feel quite uncomfortable in jeans and tee shirt.

After much debate, most agreed that one ought to dress with some formality initially, and then follow whatever appears to be the custom on the client's premises. That sounds like a sound procedure to me.

Communication

Obviously there must be communication between you and your client, and that may mean between you and many members of the client's staff. Some of that communication may be in writing, but probably most of it will be verbal or vocal communication. This is an area where it can be surprisingly easy to give offense without meaning to. Your client will expect clear and unequivocal responses to his or her inquiries, for one. Aside from the fact that your professionalism and competence are often judged on the basis of how successfully you communicate your ideas, courtesy requires that you be at pains to deliver your information so that the client can appreciate it completely without being forced to press you for elaboration. Unclear communication may be interpreted by a client to be reluctance to furnish explanations and therefore deliberate withholding of facts the client feels entitled to know. It may also be interpreted as evidence that you are not fully competent and don't really know your job well enough. (Several independent computer-services contractors I interviewed made a point of this, stressing the need to learn how to communicate well with clients.) For that reason, you must be especially careful to do whatever is necessary to communicate clearly. Accept communication as your responsibility. A wise editor put it to me many years ago about writing: "It is not enough to write so that you can be understood; you must write so that you cannot be misunderstood." That injunction should apply to speech, as well as to writing.

Trade Jargon: Its Use and Misuse

Every profession, craft and trade develops its own special jargon, and the use of that jargon becomes almost automatic among the practitioners of those professions, crafts and trades. Often, some of those terms reach the ears of the public at large and become familiar to many lay people, even if they do not fully appreciate what the terms mean. For example, now that there is a computer on virtually every desk, most of us have learned the terms *byte* and *modem,* among many other technical terms that have become prominent in everyday usage, and we may even know vaguely what those terms mean. We would probably be much less familiar with a simms memory chip and might not know a cpu from a relational database, and most clients are not seeking a technical education when they hire you to help them get something done.

Trade jargon is not confined to highly technical subjects. The client who hires you to help put together a direct response marketing campaign may very well be mystified by your references to "rolling out" and developing "response devices" if he or she is not experienced in that type of marketing. Using the more obscure jargon of your special field in talking with your client can be irritating to the client. He or she may be embarrassed to ask you to explain those terms, but is no less distressed by being unable to follow just what you are saying and suffering on frustrated ignorance or forced to plead with you for explanations.

The general rule you should follow is to minimize your use of technical jargon or, if you must use it or you use it unconsciously, explain it as you go. Most terms have simple counterparts in the English language. No one is injured or offended if you explain what they already know, but the reverse of that is not true. It is a mistake to believe that you somehow enhance your professional image by using esoteric language and shutting your clients out.

Don't Make a Mystique of Your Work Specialty

The trap that some independent contractors fall into results from a rather common fear—almost a paranoia—that allowing a client to understand what they do is harmful to their business. Their reasoning is that every client is trying to learn the contractor's trade and professional secrets so that he or she (the client) will not need to hire the contractor in the future, but will be able to do the work himself or herself. Some who write and train others in independent consulting and contracting even make a point of including in their materials special instruction on how to avoid giving away too much information to the client.

If it is ever true that clients want to learn how to do the job without your help, it is probably a most exceptional case. But even if it were true, it would be completely unethical, in my view, to deliberately withhold information from a client who has hired you to do a job. Suppose, for example, a client hires me to write a proposal and asks me such questions as, "What is a 'worry item' and how do I find and use one?" Or suppose the client asks me, "How did you arrive at the strategy you recommend?" Is the client entitled to frank and complete answers?

I think the answer is "yes." I believe the client is entitled to a complete explanation, even if it does appear that it may cost me the next proposal assignment with that client. (My experience does not, however, support that notion.) I believe that the client is entitled to any information relevant to the work I was hired for and am paid to do, and I believe that it is unethical for me to withhold that information. (Actually, it is a bad idea for a number of other reasons, but we are considering ethics and client relationships in this chapter.)

Working Relationships

There are relationships with your clients at more than one level and of more than one kind. The business relationship is that of buyer and seller or contractor and client, and may be between you and the client's comptroller or business manager, with respect to the contract and such administrative matters as billing and time records. These are almost arms-length relationships, with minimal personal contact. What we might term "working relationships" is quite another matter. These are daily, ongoing and are often between you and everyone you must relate to or work with on the client's staff, whether it is one of the client's top executives or lower level workers.

In many circumstances it can be tricky to handle these relationships just right if your work brings you into close contact with the client—i.e., with the client's staff—over an extended period. Hazards may and often do arise spontaneously out of and because of extended contact with the client and the client's staff, where they would probably not arise out of brief and intermittent contacts. This is especially the case if you do much of the work on the client's premises. That will soon be evident, even without specific scrutiny, as we examine some "dos" and "don'ts" offered here as general suggestions of ways to conduct yourself with the client and staff. Bear in mind, in reviewing these, that when a client's staff employees view an independent contractor doing the same kind of work they do on the same premises, their suspicions are often aroused that this reflects their own management's distrust, and they often come to regard the independent contractor as the enemy. Some of the suggestions made here may appear to lean over backward to avoid this perception. Indeed, that is their intent. You may not be able to combat or nullify the staff employees' suspicions and resentment, but you can take steps to avoid fostering or nourishing it.

Civility, But Not Familiarity

Do be pleasant and courteous at all times, but stop short of becoming so friendly as to be familiar and take liberties. Don't become too friendly with those working for the client. It almost surely will be resented by at least some of the staff—you will probably be regarded as an outsider by some, if not all the client's staff—and quite possibly even by the executive who contracted with you for your services. Certainly, he or she will not welcome any action on your part that may add to his or her troubles.

On one occasion, when I was an independent contractor working on site at what was then Remington Rand at Blue Bell, Pennsylvania, I quite unintentionally—thoughtlessly, to be frank—offended a senior staff member there, who interpreted a remark I made as a direct insult. I managed to patch things up and win his good will, but it took several months of conscious and patient effort to do so.

Avoid Fractious Discussions

Subjects that are most likely to lead to arguments and lost tempers are religion and politics. By all means, avoid both. Talk about baseball, computers, even your in-laws (if absolutely necessary), but not welfare programs, elections, good and bad presidents and political parties. These and other such controversial subjects can too easily turn explosive because opinions in these areas are usually based largely on emotions, rather than reason. Reason will never prevail against emotional bias.

Avoid Dogmatic and Authoritarian Pronouncements

It is, unfortunately, possible for a discussion on any subject to turn into bitter argument, especially when someone makes dogmatic declarations and proclaims his or her authority as proof of rightness. Even normally mild-mannered, easygoing individuals can become aroused to indignation and anger by such tactics, which tend to be insulting, implying that anyone who disagrees must be a boob.

The best way to avoid allowing yourself to become so embroiled is to withdraw quickly from a discussion in which one of the parties adopts the authoritarian position that all but challenges anyone and everyone to dispute a point. Don't allow an argument to develop. But above all, don't take that bombastic position yourself.

Another condition that inspires violent argument is uninformed stubbornness. Someone who is truly uninformed will sometimes make a statement that is ludicrously wrong and will resist all logical rebuttal. Moreover, the more effective the rebuttal logically, the more stubbornly the uninformed party resists, finally growing angry as the discussion grows more and more polarized.

You can deal effectively with such situations only by recognizing up front the futility of trying to change the other's mind. You cannot do it. At least, you cannot do it at the moment, and you certainly cannot do it by confrontation and argument. If you can do it at all, you can do it only by cooling down the discussion as quickly as possible and allowing the adamant party to save face by appearing to have won the argument. Given time to think calmly, without having been forced to extremes to defend his or her position, many individuals reconsider the matter and come to terms with truth. But even if they do not, you have forestalled bitterness and recrimination, and that is victory enough. There wasn't much to win anyway, was there?

Maintain the Professional Attitude

Remember always that your relationship with the client is a professional and a business relationship, and you will minimize the probability of trouble by maintaining a professional and businesslike attitude at all times. Don't be a gregarious mixer, even if it is normally your nature to be jovial and function in that mode. Try to be friendly, but in a quiet and reserved way.

Don't Try To Be a Comedian

It's a mistake to be the life of the party, especially if you encounter some of the staff at lunch or during a coffee break, and start telling funny stories. You never know when what you believe is a harmless joke someone finds offensive. Or when someone will interpret your telling funny stories as an intrusion and unwarranted presumption on your part that you can join their social circle. Let someone else be the master of ceremonies and that good fellow well met.

Don't Find a Buddy on the Job

If you work with many people on a long-term job, you almost surely will meet some you get along with especially well. During an extended working relationship with some of the client's staff, you may very well find someone with whom you have much in common, and the two of you find that you enjoy each other's company. It is hazardous and probably a mistake to allow that to develop into a close relationship. Again, it is the kind of thing that often arouses the resentment of others. Try to be as even-handed as possible in your relationships with everyone. Especially don't see any of the client's staff socially on the outside during the period of your contract.

Don't Find Romance on the Job

Even worse than finding a friendly companion among the client's staff is finding romance there. Such things as developing romantic attachments in an office environment do happen often enough, of course, and it can be truly disastrous to your relationships with the client. Even worse, it can become the subject of scandalous gossip and knowledge of it spread to other places, to the distinct detriment of your reputation and business practice. If something of the sort develops between you and someone on the client's staff, save it, at least until the job is over and done with. Postpone following it up until then.

Especially disastrous in that context, is getting more than merely passingly friendly with a client's staffer of the opposite sex who happens to be in a sensitive position—e.g., a secretary or accountant—and thus with daily access to some of the client's most sensitive and confidential business data. Should an attachment of this sort develop, it almost surely will become apparent to the client, and you may be reasonably sure that it will result in some reaction by the client, probably not a reaction you will welcome. It is not enough to be discreet here; you must avoid allowing such a situation to develop.

The Mores of Subcontracting and Temping

As I pointed out in beginning this chapter, ethical codes and client relationships are not a separate subject, important enough to merit a chapter and special discussions. That is a truth with regard to one of the most elemental factors in contracting: the contracts themselves. Many of the clauses in the contracts you sign will govern certain potential acts that ought to be regarded as moral obligations, even if they are not specified by contract, because they are principles of conduct commonly accepted by most clients and contractors. You will therefore recognize some of the situations described here because they were referred to in earlier discussions of contracts and their clauses.

One place where some of these concerns come into play most frequently is in the case of subcontracting, when a prime contractor technically is your client, but the work you do is for and brings you into close and frequent direct contact with the prime contractor's client. (Such prime contractors are often known as brokers because they don't do any of the work themselves, but fill every working slot by subcontracting with an independent contractor, so that they are really "liaisons.") That is a most common situation in many kinds of contracting and subcontracting, including engineering, drafting and computer software development, among others. In discussing these situations, when I use the word *client*, I am referring to the prime contractor's client with whom you are working directly.

Another situation giving rise to the same kind of ethical problem is that of temping. As mentioned earlier, the IRS often refuses independent business status to the independent contractor who works for a lengthy term on a client's premises, and especially when a third party is involved—i.e., client, prime contractor and subcontractor. In these cases, the IRS insists that he or she is a temporary employee, working under a W2, rather than a form 1099. Prime contractors are thus forced to make employees of the independent contractors and assign them to the client's premises. Such prime contractors are known generally as "job shops," although they like to refer to themselves as "labor contractors" or "staffing services." In either case, when working in this manner you find yourself working on the premises of a client who is not your client and not your employer. In time, as you get acquainted with the client, and here I refer to the decision-making executive who is in charge of you and what you are doing, you may even become friendly enough to have a cup of coffee together or have lengthy chats.

It may occur to one or the other of you, especially as the contract period is about to end, that it may be in your mutual interest to consider continuing to work together, but under another arrangement. The client may approach you or you may approach the client to discuss a direct contract or a staff position.

These are normally outside the pale of proper behavior, and enjoined against quite commonly by contract clauses, as discussed earlier. You probably signed a contract or employment agreement containing clauses in which you agreed to neither solicit nor accept direct contracts or employment with the client of the prime contractor or job shop for some period of time, probably one year, although many will try to make the time period two years, and many independent contractors refuse to agree to more than six months hands-off time.

It is possible that the client has signed an agreement to this effect with the prime contractor or job shop. As an alternative, the client may have agreed to pay a placement fee to the prime contractor if you accept employment with the client.

In that latter case, there is, of course, nothing wrong with accepting the client's offer, since the prime contractor has agreed to it for a consideration. Only under such a circumstance is it permissible for you to make a direct deal with the client before the stipulated period has expired.

It is conceivable, although unlikely, that neither you nor the client has signed an agreement to refrain from doing any kind of business together other than under agreement with the prime contractor. That would relieve you of legal responsibility to refrain from working directly for the client, but it would not relieve you from the moral and ethical obligation to refrain from doing so for at least six months.

Appendix:
Other Resources

Following are listings of other resources, most of which I have used and continue to use as the occasion demands. The lists start with the titles of a few books I have used, keep on my shelves for reference and recommend to you for the same purpose, and goes on to reference other materials and sources of support.

Bibliography

Arth, Marvin, and Helen Ashmore. *The Newsletter Editor's Desk Book.* 3d ed. Shawnee Mission: Parkway Press, 1984.

Bangs, David H., Jr. *The Business Planning Guide: Creating a Plan for Success in Your Own Business.* 6th ed. Dover, N.H.: Upstart Publishing Co, 1992.

Bly, Robert W. *Create the Perfect Sales Piece.* New York: John Wiley, 1985.

Brake, Barbara. *Homemade Money.* 5th ed. Cincinnati: Betterway Publications, 1994.

Broudy, Eve. *Professional Temping.* New York: Collier Books, 1989. Burgett, Gordon, and Mike Frank. *Speaking for Money.* Carpinteria, Calif.: Communication Unlimited, 1985.

Cohen, William A. *The Entrepreneur and Small Business Problem Solver: An Encyclopedic Reference and Guide.* 2nd ed. New York: John Wiley and Sons, 1990.

Edwards, Paul, and Sarah Edwards. *Working from Home.* 3d ed. Los Angeles: Jeremy P. Tarcher, 1990.

Glossbrenner, Alfred. *How To Get Free Software.* New York: St. Martin's Press, 1984.

———*The Complete Handbook of Personal Computer Communications.* 3d ed. New York: St. Martin's Press, 1989.

Holtz, Herman. *How To Succeed as an Independent Consultant.* 3d ed. New York: John Wiley and Sons, 1993.

———*The Consultant's Guide To Winning Clients.* New York: John Wiley and Sons, 1988.

———*The Consultant's Guide To Proposal Writing.* 2d ed. New York: John Wiley and Sons, 1990.

————*How To Start and Run a Writing & Editing Business,* New York: John Wiley and Sons, 1992.

————*Speaking for Profit.* New York: John Wiley and Sons, 1987.

————*Databased Marketing.* New York: John Wiley and Sons, 1992.

————*The Direct Marketer's Workbook.* New York: John Wiley and Sons, 1986.

————*Secrets of Practical Marketing for Small Business,* Englewood Cliffs, N.J.: Prentice-Hall, 1983.

Hyypia, Erik, and the editors of *Income Opportunities. Crafting the Successful Business Plan.* Englewood Cliffs, N.J.: Prentice-Hall, 1992.

Kamoroff, Bernard. *Small Time Operator.* Laytonville, Calif.: Bell Springs, 1992.

Kravitt, Gregory I. *Creating a Winning Business Plan.* Chicago: Probus, 1993.

Lewis, Herschell Gordon. *Direct Mail Copy That Sells.* Englewood Cliffs, N.J.: Prentice-Hall, 1986

Mancuso, Joseph R. *How To Write a Winning Business Plan.* Englewood Cliffs, N.J.: Prentice-Hall, 1985.

McKeever, Mike. *How To Write a Business Plan.* Berkeley, Calif.: Nolo Press, 1992.

Ries, Al, and Jack Trout. *The 22 Immutable Laws of Marketing.* New York: Harper Business, 1993.

Schectman, Gilbert. *Getting Down to Cases: Scenarios for Report Writing.* Englewood Cliffs, N.J.: Prentice Hall, 1992.

Siegel, Eric S., Brian R. Ford and Jay M. Bornstein. *The Ernst and Young Business Plan Guide.* 2d ed. New York: John Wiley and Sons, 1993.

Varner, Iris I. *Contemporary Business Report Writing.* 2d ed. Chicago: Dryden Press, 1991.

Wilder, Claudyne. *The Presentations Kit.* New York: John Wiley and Sons, 1990.

Useful Periodicals

Barbara Brabec's Self-Employment Survival Letter, Barbara Brabec. Naperville, IL 60167

DM News, 19 West 21st Street, 8th Floor, New York, NY 10010. A weekly tabloid on direct marketing.

Entrepreneur, Entrepreneur Group, Inc., 2392 Morse Avenue, Irvine, CA 92714. A thick periodical, full of business ideas.

Home Office Computing, Scholastic, Inc., 411 Lafayette St., New York, NY 10003. A most directly relevant magazine, found on newsstands.

Income Opportunities, IO Publications, Inc., 1500 Broadway, New York, NY 10036. A magazine for "opportunity seekers," with many articles and advertisements of interest to those seeking to build a home business. Found on newsstands.

Sharing Ideas!, published by Dottie Walters, P.O. Box 1120, Glendora, CA 91740. A bimonthly periodical that has become almost the bible of the public-speaking industry, but is also of interest to writers and consultants.

Target Marketing, 401 N. Broad Street, Philadelphia, PA 19108. A monthly slick paper trade magazine for direct marketers.

Mailing List Brokers

There is an abundance of mailing list brokers from whom you can rent and in some cases buy mailing lists. You can find them listed in the Yellow Pages, as well as in many other media. Here are a just a few of them, many of them branch offices of mailing list firms:

American List Counsel, Inc,
88 Orchard Road,
Princeton, NJ 08543
908-874-4300

AZ Marketing Services, Inc.
31 River Road
Cos Cob, CT 06807
203-629-8088

Woodruff-Stevens & Associates
345 Park Avenue South
New York, NY 10010

Direct Media, Inc.
200 Pemberwick Road
Greenwich, CT 06830
203-532-1000

Listworks Corp.
One Campus Drive
Pleasantville, NY 10570
914-769-7100

Qualified Lists Corp.
135 Bedford Road
Armonk, NY 10504

Worldata
5200 Town Center Circle
Suite 550
Boca Raton, FL 33486
407-393-8200

The Coolidge Company
25 West 43rd Street
New York, NY 10036
212-302-7760

Jami Marketing Services
2 Bluehill Plaza
Pearl River, NY 10965
914-620-0700

Allmedia, Inc.
4965 Preston Park Blvd.
Plano, TX 75093
214-985-4060

Dependable Lists, Inc.
950 S. 25th Avenue
Bellwood, IL 60104
708-544-1000

A Few Writing Tips

Here is a brief checklist to refer to when preparing writing marketing materials:

- Always make things as easy as possible for the client:
  - Use short words, short sentences, short paragraphs. One thought in a sentence, one subject and one main point in a paragraph. (Be sure first that you have thought the subject out and have a main point.)
  - Make it easy for the client to order, request more information or otherwise express interest by providing a return card, telephone number or other convenient way to respond.
  - Make it easy for the client to understand what you want him or her to do by explaining what to do. A great many sales are lost by advertisers who fail to tell

the customers what they want the customers to do—e.g., "Just fill out the enclosed card. . ." may sound foolish, but experience proves it works.

- A direct mail cliche which is nonetheless true, is "The more you tell the more you sell." Don't stint on copy. Include a letter, a brochure or flier of some sort and an envelope order form as a minimum. Three-quarters of the response results from the letter, and a good circular or brochure increases response by as much as one-third. (Note how many contest forms require the respondent to remove seals from one place and stick them in another place. This is part of getting prospects directly involved and so arousing their interest.)

- Don't tell it all in the letter. Divide the copy up among the various enclosures or provide additional details in the various enclosures. Make it clear that additional information and details are to be found elsewhere in the enclosures. Give the reader good reason—inducements—to read everything, if you want maximum impact.

- Geography makes a difference. Prospects who are nearby tend to respond better than those at a distance. Know your nearby ZIP codes and use them. But do test, for there are always exceptions. For example when it comes to consulting and speaking services, there is some appeal, even a kind of mystique, to the expert from a distant place, especially if you are mailing from a major industrial or business center, such as New York, Chicago, Washington or another major metropolitan area. If you are, take advantage of it, somehow, by giving it prominence in your copy. If you use envelope copy—advertising and sales messages on the outside of the envelope—do two things:

  - Use both sides of the envelope. If you are going to make a bulletin board of the envelope, you might as well get full use of it; copy on both sides pulls better than copy on one side only—if the copy is powerful.

  - Now that you've served notice that the envelope contains advertising matter, why pay first-class postage? You might as well save money by using bulk mail or, at least, something less expensive than first class.

Business Resources

- Dr. Bill McMillan, Director of Economic Development, Truman College, 1145 W. Wilson Avenue, Chicago, IL 60640, 312-907-4449.

- Service Corps of Retired Executives (SCORE), 409 Third St., SW, 4th Floor, Washington, DC 20024, 202-205-6762. (Small Business Administration–sponsored volunteer consultants.)

- Small Business Administration, 1441 L Street, NW, Washington, DC 20416 (headquarters of about 80 district offices, distributed across major cities.) SBA now has an electronic bulletin board service, SBA Online, 1-800-859-4636, and invites everyone to call and take advantage of SBA services.

- National Association of Home Based Businesses, 10451 Mill Run Cir., Suite 400, Owings Mills, MD 21117.

- National Association for the Self-Employed, 2316 Gravel Road, Ft. Worth, TX 76118.

People and Organizations in Public Speaking

- National Speakers Association (NSA), 5201 N. 7th Street, Suite 200, Phoenix, AZ 85014.
- International Platform Association (IPA), 2564 Berkshire Road, Cleveland Heights, OH 44106.
- Toastmasters International, Inc., POB 10400, Santa Ana, CA 92711.

Labor Contractors

Many independent contractors accept assignments through intermediaries that provide technical and professional temporaries, acting as brokers or job shops. There are hundreds of such firms, many of whom advertise widely in the help-wanted columns. The following is a brief starter list of such brokers, who are national companies in that they have many offices in various locations. (Some of the following also have offices listed in the special list that follows the first list.)

Consultants and Designers
960 Holmdel Rd.
Holmdel, NJ 07733
908-946-3900

Tad Resources International, Inc.
639 Massachusetts Avenue
Cambridge, MA 02139
617-868-1650

Day and Zimmerman, Inc.
1818 Market Street
Philadelphia, PA 19103
215-299-8000

Volt Information Sciences, Inc.
1221 Avenue of the Americas
47th Floor
New York, NY 10020
212-704-2400

For Retirees and Older Workers

The National Association of Temporary Services (NATS) furnishes a list of offices that deal exclusively in placing seniors in temporary positions. Some of them work with all kinds of workers in terms of skills, while others work only with one or two. The following were selected from that list as those who place technical and professional people. They are all offices of companies that are members of NATS. The companies to which they belong are all national companies, which means that each has at least ten offices operating in at least ten different states. A contact individual is identified for each.

Although the following are identified as offices that specialize in placing older workers, that does not mean that other offices also will not place older workers. Note that each of these offices is only one of many operated by company. Don't hesitate to get in touch with any of the other offices of the company. You are likely to find other offices of the company listed in your own Yellow Pages directory, but if none of the following are in a location that is convenient for you, don't hesitate to call any of these and ask for directions to other offices of their company closer to you. It is a perfectly reasonable and sensible thing to do.

Robert Half International Inc.
2884 Sand Hill Rd., Suite 200
Menlo Park, CA 94025
415-854-9700; 201 offices
Contact Max Messmer

Adia Personnel Services
100 Redwood Shores Pkwy.
Redwood City, CA 94065
415-610-1000; 750 offices
Contact Walter Macauley

Career Horizons, Inc.
695 East Main Street
Financial Centre, Stamford, CT 06901
203-975-8001; 166 offices
Contact Joel B. Miller

Dunhill Temporary Systems
P.O. Box 137
Paoli, PA 19301-0237
516-364-8800; 51 offices
Contact Howard Scott

Express Temporary Services
6300 Northwest Expressway
Oklahoma City, OK 73132
405-840-5000; 130 offices
Contact Bob Funk

Kelly Services, Inc.
999 West Big Beaver Road
Troy, MI 48084
313-362-4444; 850 offices
Contact Terance E. Adderly

Manpower, Inc.
5301 N. Ironwood Road
P.O. Box 2053
Milwaukee, WI 53201
414-961-1000; 820 offices
Contact Mitchell Fromstein

Office Specialists
Corporate Place,
128 Audubon Rd., Building 1, #3
Wakefield, MA 01880
617-246-4900; 50 offices
Contact Robert M. Whalen

The Olsten Corporation
One Merrick Avenue
Westbury, NY 11590
516-832-8200; 500 offices
Contact Frank Liguori

Remedy Temporary Services
32122 Camino Capistrano
San Juan Capistrano, CA 92675
714-661-1211; 40 offices
Contact Robert McDonough

Salem Services, Inc.
1333 Butterfield Road
Downers Grove, IL 60515
708-515-0500; 36 offices
Contact J. Marshall Seelander

Snelling & Snelling, Inc.
12801 N. Central Expressway
Suite 700
Dallas, TX 75243
214-239-7575; 87 offices
Contact Brian Dailey

Talent Tree, Inc.
9703 Richmond Avenue
Houston, TX 77042
713-974-6509; 104 offices
Contact Mike Willis

Uniforce Services
1335 Jericho Turnpike
New Hyde Park, NY 11040
516-437-3300; 81 offices
Contact Rosemary Maniscalco

Volt Temporary Services
2401 N. Glassell Street
P.O. Box 13500
Orange, CA 92665
714-921-8800; 83 offices
Contact Mary Smith

Seminar Tips

Seminars can be among the most important tools available to you for earning income and marketing your services. Producing and presenting seminars therefore deserves serious consideration. With that in mind and for that reason, a few tips and reminders are offered here for ready reference:

- A typical day's session runs approximately six hours, three in the morning and three in the afternoon, with midmorning and midafternoon breaks of about ten minutes. The lunch break is thus usually from 90 minutes to two hours.
- Back-of-the-room sales may be conducted prior to the start of the morning session, during the lunch break and after the close of the afternoon session. They should not be conducted during the presentations.
- Whether you sell books, tapes or other materials, you should include handout materials as part of the seminar. These are often a major inducement to registration and attendance. It helps, from that viewpoint, if they are available only to registered attendees.
- In many cases you can get excellent materials free of charge from government agencies, associations, community groups, large corporations, schools and other sources. You may also be able to get speakers free of charge from such sources. Computer consultants should always consider asking local computer dealers and others in the computer business for any such support—that is, speakers and handout materials. (By *government agencies,* I mean federal, state and local governments.)
- Visual aids are often available from such sources as those just named and include movies, slides, film strips, transparencies and posters.
- One profitable idea is the mini-seminar. I have held two- to three-hour sessions for 10 to 15 people in my offices on Saturday mornings, charging what was even then a rather modest fee—$25. If your offices are at home and unsuitable for the purpose, you usually can arrange to rent an inexpensive room somewhere.
- A variant on the mini-seminar idea is to conduct two such sessions, one in the morning and another in the afternoon, enabling yourself to register twice the number of attendees for a single day. The total cost is only slightly greater for two sessions, and the income is, of course, thus doubled.

Proposal Tips

Nothing is more important to the marketing of your professional services than the proposal, used properly and with a few tips in mind:

- Analyze the client's problem closely and be sure you understand it fully before committing yourself to a plan of action. Don't rush to the word processor before you have done so.
- Devise a specific strategy upon which to base your proposal.
- Be specific in what you propose to do and furnish to the client.
- Avoid hyperbole. Stick with nouns and verbs. Use adjectives and adverbs sparingly. Avoid superlatives.

- Quantify as much as possible in providing details.
- Avoid potential disputes by being sure to specify exactly what your quoted cost estimate covers.

A Suggested Proposal Outline

A proposal usually includes four chapters or sections as well as front matter in the formal proposal and back matter in many cases.

This format may, of course, be modified to suit your own preferences, the circumstances or the dictates of the client. An informal or letter proposal would not have front and back matter or separate chapters, although it would follow the general philosophy of this outline. Some requests for proposals (RFPs) mandate a proposal format, and some companies have a standard format specified for their proposal, either of which you will of course follow, in such case.

Front Matter

- Copy of transmittal letter
- Executive summary: Provide an abstract of most important points that demonstrate your best arguments.
- Table of Contents

Section/Chapter I: Introduction

- About the offeror: Briefly introduce your firm, sketch your company and qualifications, refer to details to be found later and make other opening statements.
- Understanding of the requirement: Make a brief statement of your understanding of the requirement, in your own language (don't echo the RFP), leave out the trivia and focus on the essence of the requirement, providing a bridge (transition) to the next chapter.

Section/Chapter II: Discussion

Discuss the requirement, analyzing, identifying problems and exploring and reviewing approaches (with pros and cons of each). Include similar discussions of all relevant matters, technical, management, schedule and other important points, including worry items. This is the key section in which to sell the proposed program, make the emotional appeals (promises), explain the superiority of the proposed program and demonstrate the validity of the proposer's grasp of the problem, of how to solve it, of how to organize the resources and otherwise *sell* the idea. This section should culminate in a clear explanation of the approach selected, bridging directly into the next chapter. Include graphics as necessary, especially a functional flowchart, explaining the approach and technical or program design strategy employed.

Section/Chapter III: Proposed Project

This is where the specifics appear—staffing and organization (with organization chart), résumés of key people, either here or later in this chapter, but at least introduced here by name. Key elements are these:

- Project management: Explain procedures, philosophy, methods and controls; relationship to parent organization and reporting order; and other information on both technical and general or administrative management of project. (This may be a separate chapter or even a separate volume, for larger projects.)
- Labor-loading: Explain major tasks and estimated hours for each principal in each task (use tabular presentation), with totals of hours for each task and totals of hours for each principal staff member.
- Deliverable items: Specify, describe and quantify, as explained.
- Schedules: Specify, as explained. (Use a milestone chart, if possible.)
- Résumés: Provide résumés of yourself and any associate or consultant you may use.

Section/Chapter IV: Company Qualifications

Describe the company, past projects (especially those similar to one under discussion), resources, history, organization, key staff, other résumés, testimonial letters, special awards and other pertinent facts.

Appendixes (Back matter)

Append detailed data, drawings, papers, bibliography, citations and other material that some, but not all, readers will want to read.

Tips on Proposal Graphics

Graphics—drawings, charts and graphs—add greatly to clear and easy communication of ideas. By all means, make use of them. With today's computers and software, anyone can prepare professional-quality graphics.

A good illustration requires little explanation, and that is the way to test the quality of any illustration: Does it require explanation, and if so, how much? Is the illustration clear or is it "clever"? Forget about clever devices and artistic considerations; the purpose of an illustration is to communicate information accurately and efficiently. If the reader has to puzzle over the meaning or study the illustration to understand it, the reader will probably set your proposal aside with a sigh and go on to the next one. The basic rule is to make it as easy as possible on the reader. Cleverness is all too often the death of meaning and understanding, and therefore the death of the sale.

For function charts, use the "why? how?" technique to generate the chart and to test it. Going from left to right (or from top to bottom, if you prefer that progression), ask why? of each box, and the answer should be in the next box. Going the other way—in reverse—ask how, and the answer should be in the next box. If the answers are not very clear, consider adding boxes for more detail or changing the wording in the boxes. Charts, like text, should go through drafts, editing, reviews and revisions.

Using Headlines, Glosses and Blurbs in Proposals

Proposals are not exciting literature, and at best are fatiguing to read in quantities, as customers are compelled to do. Anything you can do to make it easier for the reader will help you, in the end, in two ways:

1. It will help you get your own messages across and pierce the consciousness of readers who may be reading mechanically and without full appreciation, by the time they get to your opus.
2. You will earn the reader's gratitude, which can do nothing but help your case.

Headlines

Use headlines—sideheads and centerheads—as freely as you can, as often as you can. Use them to summarize messages, to telegraph what a paragraph or page is about, what the main message is. But use them also to *sell*. That is, use the headline to summarize promises—benefits—and proofs. Use them to remind the reader of the benefits and reinforce the proofs.

Glosses

A gloss is a little abstract in the margin of a page that summarizes the text next to it. Usually, there is at least one gloss on a page, and often there are several. Like headlines, glosses can and should be used to help sell the proposal by focusing on benefits and proofs.

Blurbs

A blurb is very much like a gloss, except that it is used less frequently and is thus somewhat broader in scope and, usually, of greater length. A blurb generally appears after a major headline (usually a center head) or chapter title. Like headlines and glosses, blurbs should be used to sell, as well as to sum up information and communicate generally.

Online Utilities and Public Databases

The following is a brief sampling of just a few of the better-known or more prominent online utilities and public databases, available on a subscription basis, usually with a "connect time" charge and a monthly minimum. The differences between the two is principally that, while the public database is strictly an information source and is usually used only by those who need the information for business and professional purposes, the online utilities offer services many use for entertainment and amusement as virtual public meeting places to seek out others with similar interests for chats and debates via the "forums" or "conferences."

Online Utilities

CompuServe Information Service, Inc.
5000 Arlington Centre Boulevard
Columbus, OH 43220
614-457-8650, 800-848-8199

Prodigy Services Company
445 Hamilton Avenue
White Plains, NY 10601
914-448-8000

GEnie
GE Information Services
401 N. Washington Street
Rockville, MD 20850
301-340-4000, 800-638-9636

Minitel USA
888 7th Ave.
28th Floor
New York, NY 10106
212-399-0080

Public Databases

Dialog Information Services, Inc.
3460 Hillview Avenue
Palo Alto, CA 94304
415-858-3792, 800-334-2564

BRS Software Products
5 Computer Drive South
Albany, NY 12205
518-446-0490, 800-235-1209

Dow Jones News Retrieval
P.O. Box 300
Princeton, NJ 08543-0300
609-452-1511, 800-522-3567

NewsNet, Inc.
945 Haverford Road
Bryn Mawr, PA 19010
215-527-8030, 800-345-1301

Mead Data Central (Nexis and Lexis)
9393 Springboro Pike
P.O. Box 933
Dayton, OH 45401
800-227-4908.

Government Electronic Bulletin Board Systems

Many bulletin board systems (BBSs) are operated in federal government offices. Some are official organs of the agencies, while others are quasi-official in that an employee operates them under the sponsorship of the agency. One that may be the most important of these, from your viewpoint as a computer consultant, is that of the General Services Administration, which has jurisdiction over computer standards-setting and other matters relating to computer procurement by government agencies. This bulletin board carries important information about requirements, procurements and major contractors, so that it can be an important source of business leads for you. The BBS number follows:

General Services Administration BBS
202-501-2014

A message on this BBS provides a list of telephone numbers for each of the 10 GSA regional contacts for ADP Technical Service Requirements contracts as follows:

- Capital Zonal Office
  202-708-5726
- Central Zonal Office
  205-895-5019
- Eastern Zonal Office
  215-656-6293
- Pacific Zonal Office
  415-744-8527
- Western Zonal Office
  817-334-3686

Another highly important BBS is run by the Small Business Administration. The main number for SBA Online was given in "Business Resources" earlier in this appendix, with a toll-free number for access at 2400 baud, but there are three numbers for SBA Online:

1. 202-205-7265
2. 800-697-4636: access at 9600 baud
3. 800-859-4636: access at 2400 baud

A few of the many other BBSs in government offices are listed here:

U.S. Navy
Navy ADA Language System:
703-604-4624

Department of Agriculture
Library: 301-504-6510
Nutrition: 301-436-5078
Library of Congress; ALIX II:
202-707-4888

Department of Commerce
Census Bureau: 301-763-4574
Geological Survey: 703-648-4168

Department of Education
202-219-2011; 202-219-2012

Department of the Interior
703-787-1181

State Department USAID/Permanent BBS
703-715-9806: 2400 baud
703-715-9851: 9600 baud

National Science Foundation SRS
(Science and Research Studies):
703-306-1234

Government Printing Office
Federal Bulletin Board:
202-512-1387

A Few Tips on Pursuing Government Contracts

Individuals can do business with federal agencies, just as major corporations do. (I have personally won and performed on many government contracts.) The advantages of doing business with the government far outweigh the drawbacks. The following suggestions will help you get started learning the ropes of selling your services to the government.

- To get on bidders lists: Ask for Standard Form 129, Application for Bidders List, and make it up in enough copies to distribute to the various agencies with whom you think you can do business.

- To learn of outstanding bid and proposal opportunities, subscribe to the *Commerce Business Daily* (CBD) or to any public database that carries CBD Online, such as CompuServe or Dialog. Also, visit government bid rooms and monitor the notices on the bulletin boards. Don't depend entirely on the Form 129.

- If you are in or near a General Services Administration Business Service Center, take the time to visit and talk to the people there. They are located in Boston, New York, Philadelphia, Washington, Atlanta, Chicago, Kansas City, Denver, Houston, Fort Worth, Los Angeles, San Francisco and Seattle. If not, write to the one in Denver—P.O. Box 25006 Denver Federal Center, Building 41, Denver, CO 80225-0006, 303-236-7329—and request all available information on doing business with the government.

- Write to the contracting offices of major agencies—e.g., the Department of Defense, NASA, Department of Health and Human Services and others—and make the same request for information.
- Watch the CBD for announcements of free seminars on competing for federal contracts. They are held frequently, sponsored by members of Congress in their districts.
- Get acquainted with the contracting officers of federal agencies near you. Personal contact helps a great deal in all marketing, and a friendly contracting officer can be very helpful.

Index